Samuel Musgrave

On the Graecian mythology

and an examination of Sir Isaac Newton's objections to the chronology of the

Olympiads

Samuel Musgrave

On the Graecian mythology
and an examination of Sir Isaac Newton's objections to the chronology of the Olympiads

ISBN/EAN: 9783742875358

Manufactured in Europe, USA, Canada, Australia, Japa

Cover: Foto ©berggeist007 / pixelio.de

Manufactured and distributed by brebook publishing software (www.brebook.com)

Samuel Musgrave

On the Graecian mythology

TWO
DISSERTATIONS.

I. ON THE
GRÆCIAN MYTHOLOGY.

II. AN
EXAMINATION

OF

Sir Isaac Newton's Objections to the
Chronology of the Olympiads.

BY THE LATE

SAMUEL MUSGRAVE, M.D. F.R.S.

LONDON,

PRINTED BY J. NICHOLS.

MDCCLXXXII.

ADVERTISEMENT.

THOUGH many of the moſt liberal Subſcribers in the following liſt have ſignified an intention of requiring only one copy, it has been thought proper to do juſtice to their generoſity, by ſetting down the full amount of each ſubſcription.

The Names of the Subscribers.

A.

Mr. Abbot, of C. C. C. Oxf. L. P.
Mr. Abbott, Stud. of Chr. Ch. Oxf. L. P.
Rev. J. Acland, Vicar of Broad Clyft, Devon. 2 L. P.
Rev. Dr. Adams, Mafter of Pemb. Coll. Ox. 2 L. P.
Rev. Mr. Adams, Rector of S. Okington, Effex.
Rev. Mr. Alderfon, Norwich, L. P.
Rev. Mr. Allenfon, M. A. Fell. of Jefus C. Camb.
Mr. Allington, of Peter-houfe, Camb.
Rev. Mr. Alt. L. P.
Lord Vifcount Althorp. L. P.
Mr. Anftey, B. A. of Trinity Coll. Camb.
R. Pepper Arden, Efq.
Rev. Mr. Atkinfon, Fell. of Queen's Coll. Oxf.
Rev. Mr. Atkinfon, M. A. Fell. of Trin. H. Camb.
John Aubrey, Efq. 2 L. P.
Mr. Auftin, Fell. of Wadham Coll. Oxf. L. P.
Mrs. Awfe of Windfcott, Devon. 10 L. P.

B.

The Lord Bagot, 2 L. P.
Rev. Dr. Bagot, Dean of Chriftchurch, Ox. 6 L. P.
Richard Bagot, Efq. 2 L. P.
Sir G. Baker, Bart. M. D. Phyfician to the Q. L. P.

Rev

Rev. Mr. Bale, Stud. of Chriftchurch, Oxf. L. P.

Sir Jofeph Banks, Bart. P. R. S. 12 L. P.

John Baring, Efq. M. P. for the c. of Exeter, 2 L.P.

Rev. Dr. Barker, Princ. of Brafennofe Coll. Ox. L.P.

Rev. Mr. Barnard, of C. C. C. Oxf. L. P.

The Hon. Daines Barrington, 2 L. P.

Rev. Dr. H. Barton, Warden of Merton Coll. L.P.

A perfon unknown, by Dr. H. Barton, 40 S. P.

Rev. Philip Barton, Subdean of Exeter, 10 L. P.

Mr. Baftard, L. P.

Rev. Mr. Bathurft, of New Coll. Oxf.

Edw. Dav. Batfon, Efq. L. P.

C. W. Batt, Efq. 2 L. P.

J. T. Batt, Efq. 2 L. P.

Mr. Thomas Bayley, of Jefus Coll. Camb.

Mr. J. Baynes, M. A. Fell. of Trin. Coll. Camb. L.P.

Rev. Mr. J. Baynes, M. A. of Queen's Coll. Ox. L.P.

Rev. Dr. Beadon, Mafter of Jefus Coll. Camb. L.P.

Henry Beavis, Efq. L. P.

Rev. Mr. Becke, of Oriel Coll. Oxf. L. P.

—— Bell, Efq. L. P.

Capt. Bellew, Exeter, L. P.

Thomas Bellew, Efq. Exeter, L. P.

Mr. T. Belton, of Great Torrington, Devon.

Mr. Benfon, of Jefus Coll. Camb.

Rev. Scrope Berdmore, B. D. Fell. of Mert. C. L.P.

Rev. Dr. Beridge, of Jefus Coll. Camb. L. P.

Rev. Dr. Bernard, late Provoft of Eton, 10 L. P.

Mr. Scrope Bernard, M. A. Stud. of C.C. Ox. L.P.

Rev. Dr. Biffet, 2 L. P.

Charles

Charles Blagden, M. D. L. P.

Alexander Blair, Efq. L. P.

Mr. Boddam, Fell. Com. Trin. Coll. Camb. L. P.

W. M. Bogdani, Efq. 2 L. P.

Mr. Bonney, B. A. of Jefus Coll. Camb.

Rev. Mr. Booth, Fell. of Merton Coll. L. P.

Mrs. Borlafe, of Cornwall. 4 L. P.

Rev. Mr. Borlafe, Regif. of the Univ. Camb. L.P.

Rev. Mr. Bowen, Fell. of Brafennofe Coll. Oxf.

Fofter Bower, Efq. 2 L. P.

Rev. Mr. Thomas Boyce.

Rev. Mr. Bradley, of C. C. C. Oxf. L. P.

Dr. Brandé. L. P.

Rev. Mr. Breeks, Fell. of Queen's Coll. Oxf.

Mr. Egerton Bridges, of Queen's Coll. Camb.

The Hon. Ch. Brodrick, of Clare H. Camb. L. P.

Robert Bromfield, M. D. F. R. S. L. P.

Rev. Dr. Brown, Mafter of Pembroke H. Camb.

Ifaac Hawkins Browne, Efq. 2 L. P.

Jacob Bryant, Efq. 2 L. P.

George Buck, Efq. L. P.

Lewis Buck, LL. D.

Mr. Buckland, of C. C. C. Oxf. L. P.

Rev. Benj. Buckler, D. D. Fell. of All S. C. L. P.

Rev. Dr. Buller, Canon of Windfor. 4 L. P.

Rev. Mr. Bulmer, of Jefus Coll. Camb.

John Burgefs, M. D 2 L. P.

Mr. T. Burgefs, of C. C. C. Oxf. L. P.

Richard Buck, jun. Efq. L. P.

Mr. Jofeph Burrow, Exeter. L. P.

Rev.

Rev. Mr. Burt, Stud. of Chr. Ch. Oxf. 2 L. P.

Francis Burton, Efq. 2 L. P.

Charles Butler, Efq. 2 L. P.

C.

Library of Catharine Hall, Cambridge. L. P.

Rev. Mr. Carr. L. P.

Rev. Mr. Chancellor Carrington. L. P.

Mrs. Cartwright, of Exeter. L. P.

Rev. Dr. Caryl, late Mafter of Jefus C. Camb. L.P.

Rev. Mr. Caufley, M. A. Fell. of Trin. Coll. Camb.

Edward Chamberlayne, Efq. 2 L. P.

Rev. Mr. Chamberlayne, Fell. of Eton Coll. L. P.

Mrs. Chambers. L. P.

Mr. Henry Chambers. L. P.

Anthony Champion, Efq. 2 L. P.

Earl of Charlemont. L. P.

Mr. Nathaniel Chauncy.

The Lord Bp. of Chefter. L. P.

Sir John Chetwood, Bart. L. P.

Rev. Dr. Chevalier, Mafter of St. John's C. Cam. L.P.

Sir John Chichefter, Bart. L. P.

John Chichefter, Efq. of Youlfton, Devon. 2 L. P.

Library of Chrift's Coll. Camb.

Rev. Mr Churchill, of C. C. C. Oxf. L. P.

Mr. Chute, jun. Fell. Comm. of Clare H. Camb.

William Clark, Efq.

Mr. Clarke, M. A. Fell. of Caius Coll. Camb.

Lieut. Hamilton Clarke, Exeter. L. P.

Rev.

Rev. Mr. Cleaver, Fell. of Brasennose Coll. L. P.

Rev. Mr. Euseby Cleaver. 2 L. P.

Mr. Coates, of C. C. C. Oxf. L. P.

Rev. W. Cole, M. A. Fell. of King's Coll. Camb.

Rev. Mr. Collier, Prof. Heb. Camb. L. P.

Rev. Mr. Collinson, Fell. of Q. Coll. Ox. L. P.

George Colman, Esq. L. P.

Rev. Dr. Colombine, Norwich. L. P.

Mr. Conant, Fell. Com. of Trin. C. Camb. L. P.

Rev. Dr. Coneybeare. 2 L. P.

Hon. and Rev. E. Conway, B. A. F. of A.S.C. L.P.

—— Conway, Esq. L. P.

Rev. Dr. Cooke, Dean of Ely L. P.

Rev. Mr. Cooke.

Rev. Dr. Cooper. 2 L. P.

Mr. Coulthurst. Fell. Com. of Peterhouse, Camb.

Peregrine Courtnay, Esq. L. P.

Rev. Mr. Cracherode, M. A. Stud. of C.C.Ox. 10L.P.

Rev. Mr. Cranke, M. A. F. of Trin. Coll. Cam. L.P.

Rev. Mr. Craven, B. D. Prof. of Arab. Camb.

Rev. Mr. Crowe, of New Coll. Oxf. L. P.

John Culme, Esq. 2 L. P.

Rev. J. Cutler, M. A. Rec. of Droxford, Ha. L.P.

Mr. Samuel Cutler. 2 L. P.

D.

Rev. Mr. D'Aeth, Rector of Eythorn, Kent.

Denis Daly, Esq. L. P.

Rev. Dr. Dampier, Preb. of Durham. 2 L. P.

Rev.

Rev. Mr. Darling.

Rev. Mr. Davidſon, F. of Peter-houſe, Camb. L.P,

Rev. Dr. Davies, Maſter of Eton-School. 2 L. P.

Rev. Mr. Davies, M. A. F. of Trin. C. Camb. L. P.

—— Dawkins, Eſq. Gent. Com. of C.C. Ox. L.P.

The Duke of Devonſhire. L. P.

The Dutcheſs of Devonſhire. L. P.

Mr. Dickenſon, of Peter-houſe, Camb. L. P.

Rev. Dr. Diſney. L. P.

Rev. Fletcher Dixon. L. P,

Rev. John Dixon.

Lieutenant Henry Dodd, Exeter. L. P.

Richard Doidge, Eſq. 4 L. P.

John Engliſh Dolben, Eſq. 2 L. P.

Rev. Mr. Donald, M. A. of Queen's Coll. Oxf.

Rev. Mr. Done, Fell. of C. C. C. Oxf. 2 L. P.

Rev. Mr. Dowſon, M. A. of Queen's Coll. Oxf.

Edward Drummond, Eſq. of Chr. Ch. Oxf, 2 L.P.

Robert Hay Drummond, Eſq. of C. C. Ox. 2 L.P.

Richard Roſe Drew, Eſq. Exeter, 2 L. P.

Dr. Duck, Norwich. L. P.

The Lord Viſcount Duncannon. L. P.

Mr. Dunſterville, Surgeon at Plymouth. L. P,

E.

Mr. C. Edmonſtone, of Chr. Ch. Oxf. L. P.

Rev. Dr. Edwards, Fell. of Jeſus Coll. Oxf. L. P.

Rev. Mr. Edwards.

Turner

Turner Edwards, Efq. of Jefus Coll. Oxf. L. P.

William Elford, Efq. Plympton. L. P.

Mr. Eliot, jun. Fell. Com. of Pemb. Hall, Camb.

Earl of Egmont. L. P.

Countefs of Egmont. L. P.

Right Hon. Welbore Ellis. 2 L. P.

Rev. Mr. Elfton.

Rev. F. H. Egerton, M. A. Fell. of All Souls Coll.

Rev. Mr. Empfon, M. A. of Cath. Hall, Camb.

Library of Eton College. 2 L. P.

Sidney Evelyn, Efq. L. P.

Mr. Everitt, of Trinity Hall, Camb.

Literary Society at Exeter.

Mr. Eyton, of Jefus Coll. Oxf. L. P.

F.

Thomas Falconer, Efq. of Chefter. L. P.

Rev. Dr. Farmer, Mafter of Emanuel C. Cam. L.P.

Mr. Farquharfon, of Peter-houfe, Camb.

Mr. Filmer, of C. C. C. Oxf. L. P.

Hon. and Rev. D. Finch, B. A. Fell. of A. S. C.

Rev. Mr. Foley, Fell. of Brafennofe Coll. L. P.

Rev. Mr. Foot, Rector of Drew, L. P.

Rev. Mr. J. Foot. L. P.

Rev. Mr. Foffe.

Rev. Dr. Fothergill, Prov. of Q. Coll. Oxf. 3 L.P.

Rev. W. Fothergill, M. A. Fell. of Q. Coll. Oxf.

Mr. T. Fothergill, M. A. of Queen's Coll. Oxf.

Mr. James Fothergill, B. A. of Queen's Coll. Ox.

John

John Fraine, Efq. Chelfea. L. P.

Capt. Fraine, Bath. L. P.

R. Frankland, Efq. of Chr. Ch. Oxf. L. P.

Rev. Mr. Fulham. L. P.

G.

Rev. Mr. Gandy. L. P.

John Whalley Gardiner, Efq.

Rev. Mr. Gardiner, Fell. of Catharine H. Camb.

Dr. Garthfhore.

Edward Gibbon, Efq. 2 L. P.

Geo. Abr. Gibbs, Efq. of Exeter. L. P.

Rev. Francis Gifborne. L. P.

Thomas Gifborne, Efq. 12 L. P.

Thomas Gifborne, M. D. 7 L. P.

The Lord Bifhop of Gloucefter. L. P.

Dr. Glynn, Fell. of King's Coll. Camb. 2 L. P.

Mr. Ambrofe Godfrey.

Rev. John Gooch, M. A. of Chr. Ch. Ox. 2 L. P.

Rev. Mr. Gould, M. A. of Cl. Hall. Camb. L. P.

Rev. Dr. Graham, of Netherby, Cumb. 2 L. P.

Mr. Edw. Granger, Exeter. L. P.

William Graves, Efq. 2 L. P.

Rev. Mr. Green, Chaplain of St. Thomafes Hofp.

Rev. Mr. Greene, Norwich. L. P.

Rev. Mr. Greenfide, Fell. of Cath. Hall. Camb.

Rev. Mr. Gratton, M. A. of Trinity Coll. Camb.

Hon. Mr. Greville. L. P.

Mrs.

Mrs. Griffith, L. P.
Mrs. Griffith of Liſſon Green. L. P.
Richard Griffith, jun. Eſq. L. P.
Miſs Griffith. L. P.
Mr. Griffith, of C. C. C. Oxf. L. P.
Mr. Chriſtopher Gullet, Exeter. L. P.
Mr. J. Gunning, Serj. Surg. Ext. to the King. L.P.

H.

Mr. Hailſtone, of Trinity Coll. Camb.
Rob. Halifax, Apoth. to the King's Houſehold. L.P.
T. Hall, Eſq. 2 L. P.
Mr. Hallifax, of Magd. Coll. Oxf. L. P.
William Hamilton, Eſq. 2 L. P.
Rev. Dr. Hey, of Sidney Coll. Camb.
Mr. Hamley, Surgeon, at Milbrook. L. P.
Joſ. Chaplin Hankey, Eſq.
Rev. Mr. Hardcaſtle, Fell. of Merton Coll. L. P.
Chriſtopher Harris, Eſq. Plymouth. L. P.
David Hartley, Eſq. L. P.
Mr. Harwood, M. A. Stud. of Chr. Ch. Oxf. L.P.
John Hatſell, Eſq. 2 L. P.
Mr. Hawkins, F. C. of Trin. Coll. Camb. L. P.
Mr. G. Hawkins, Surg. to the K's Houſhold. L.P.
Iſaac Hawkins, Eſq. 2 L. P.
Rev. E. Hawtrey, M. A. Rect. of Monxton, Ha. L.P.
Mr. Haydon, Bookſeller at Plymouth. 2 S. P.
Mr. Haye, B. A. Stud. of Chr. Ch. Oxf. L. P.

Mr.

Mr. T. Haye, B. A. Stud. of Chr. Ch. Oxf. L. P.

The Hon. Mr. Juftice Heath. 10 L. P.

Rev. Mr. Heath, Mafter of Harrow School. 2 L.P.

Rev. Mr. Heath, of Eton. L. P.

William Heberden, M. D. 10 L. P.

Rev. Mr. Heberden, Prebendary of Exeter.

Rev. Dr. Hemington, Canon of Chr. Ch. Ox. 2L.P.

Rev. Mr. Henley. L. P.

Mifs Henfhaw. L. P.

Bold Fleetwood Hefketh, Efq. of Magd. C. Oxf.
1 L. and 1 S. P.

Rev. Mr. Hill, Taviftock.

Mr. Hill, of C. C. C. Oxf. L. P.

Mr. James Hine, Exeter. L. P.

Rev. Mr. William Hole.

Rev. Mr. Hole, Archdeacon of Barnftaple. 2 S. P.

Rev. R. Hole, Rector of N. Tawton, Devon.

Rev. Dr. Hollingberry, Archd. of Chichefter. L.P.

Rev. T. Hornfby, Sav. Prof. of Aftron. Oxf. 4 L. P.

Mr. J. W. Hofkins, of Magd. Coll. Oxf. L. P.

Mr. How, of Peter-houfe, Camb.

Mr. Hubberfty, B. A. Fell. of Queen's C. Camb.

Rev. Mr. Hughes, Chaplain of the D. Y. Plym. L.P.

W. Hunter, M. D. Phyfician Ext. to the Q. 20L.P.

Rev. Mr. Hume, of Weftm. School. 2 L. P.

I.

Rev. Dr. Jackfon, Canon of Chr. Ch. Oxf. 4 L.P.

Rev. Mr. W. Jackfon, Stud. of Chr.Ch. Oxf. 2L.P.

Gregory Jackfon, Efq. Exeter. L. P.

Rev.

Rev. Mr. Jackſon, of Hertf. Coll. Oxf.
Miſs James. L. P.
Sir Richard Jebb, Bart. M. D. 2 L. P.
Benj. Incledon, Eſq. 1 L. and 4 S. P.
Joſ. Ingram, Eſq. M. A. Fell. of A. S. C. L. P.
R. P. Jodrell, Eſq.
Rev. S. Johnes, M. A. Fell. of All Souls Coll.
Dr. Samuel Johnſon. L. P.
Alex. Johnſon, M. D. L. P.
Library of St. John's Coll. Camb. L. P.
William Jones, Eſq. L. P.
Rev. Dr. Jubb, Canon of Chr. Ch. Oxf. 2 L. P.

K.

Mr. Keeble, of C. C. C. Oxf.
Dr. William Keir, Hatton Street.
Rev. Dr. Kennicott, Canon of Chr. Ch. Oxf. 2L.P.
Mr. Lewis Ker, M. B.
Mr. S. Kilner, M. A. Fell. of Merton Coll. L. P.
Library of King's Coll. Camb. L. P.
Henry Kitſon, Eſq. Exeter. 2 L. P.
Mr. Knipe.
Francis Knox, Eſq. L. P.

L.

Mr. Lambard, M. A. Stud. of Chr. Ch. Oxf. L. P.
Mr. T. Lambard, B. A. Stud. of C. C. Oxf. L. P.
Rev. Mr. Lambert, Fell- of Trin. Coll. Camb.
The Lord Biſhop of Landaff. 10 L. P.

Rev.

Rev. W. Langford, D.D. UnderMaf. of EtonSc. L.P.
Sir James Langham, Bart.   10 L. P.
Chriftopher Langlois, Efq. 2 L. P.
Bennet Langton, Efq. 2 L. P.
Mr. Laurence, of C. C. C. Oxf.   L. P.
Rev. Mr. Law, Archdeacon of Carlifle. L. P.
Edward Law, Efq. M. A. Fell. of St. Peter's C. Ca.
Ewin Law, Efq.
Rev. Dr. Lee, Warden of Winchefter Coll. L. P.
Hon. W. Legge, M. A. Fell. of A. S. C. Ox. L.P.
The Lord Vifcount Lewiffham.   2 L. P.
Rev. Mr. Leigh, Norwich.   L. P.
Literary Society at Lincoln.   L. P.
The Lord Biffop of Litchfield.   2 L. P.
Rev. Dr. Lloyd, Dean of Norwich. L P.
Rev. Dr. Lort, F. R. S. and A. S.   L. P.
Mr. Lovering, of St. John's Coll. Camb.
James Luke, Efq. Exeter.   L. P.

M.

Edmund Malone, Efq.   L. P.
Mr. Manley, Plymouth.   L. P.
Dr. Manning, Norwich.   L. P.
Mr. Manfell, M. A. Fell. of Trin. Coll. Camb.
Rev. Mr. Mantell, M. A. Fell. of Benet Coll. Camb.
Rev. Mr. Marfhail, Maf. of the Free Sch. Ex. L. P.
Samuel Martin, Efq. 2 L. P.

Rev. Mr. Maffingberd, of Magd. Coll. Oxf. 1 L.
and 1 S. P.

Rev. Mr. W. Maffingberd, of Magd. C. Oxf. L.P.

Mr. F. Maffingberd, of Hertford Coll. Oxf.

Mr. Mathew, LL. B. Fell. of Jefus Coll. Camb.

Mr. Mathias, M. A. Fell. of Trin. Coll. Camb.

Rev. Mr. Emanuel May.

Lieut. General Melvill. L. P.

John Merivale, Efq. of Exeter. L. P.

Rev. Mr. Metcalfe, M. A. Fell. of Chrift's C. Cam.

Rev. Dr. Milles, Dean of Exeter. 2 L. P.

Thomas Milles, Efq. LL. B. Fell. of A. S. C. L. P.

Francis Milman, M. D. L. P.

Mr. Moneypenny, of Peter-houfe, Camb. L. P.

Mr. Moneypenny, jun. of Peter-houfe, Cam. L.P.

Rev. Dr. Monkhoufe, Fell. of Queen's C. Ox. L.P.

Donald Monro, M. D. L. P.

Fred. Montagu, Efq. L. P.

Rev. J. Montagu, M. A. Fell. of All Soul's Coll.

Rev. G. Moore, Canon Refidentiary of Ex. 10 L.P.

Rev. T. Moore, Vicar of St. Veryan, Corn. L. P.

Earl of Mornington, Stud. of Chr. Ch. Oxf. L. P.

Rev. Mr. Morrice, Stud. of Chr. Ch. Oxf. L. P.

Rev. Hooper Morrifon, Rec. of Atherington. L. P.

Cha. Morton, M. D. Princ. Librar. Mufeum. 4L.P.

Mr. C. Mofs, Stud. of Chr. Ch. Oxf. L. P.

Rev. Mr. Mounfey, M. A. Fell. of Jef. Coll. Camb.

Mr. Tho. Mudge, Plymouth. L. P.

Mr. J. Mudge, Surgeon at Plymouth. L. P.

W. H. Mugglefione, M. B. L. P.

William

William Muggleſtone, Eſq. L. P.
Rev. Mr. Murthwaite, Fell. of Queen's Coll. Oxf.
Joſeph Muſgrave, Eſq. L. P.
George Muſgrave, Eſq. L. P.
R. Myddelton, Eſq. Gent. Com. of C. C. Ox. L.P.
W. Myddelton, Eſq.

N.

Mr. Nation, Exeter. L. P.
Chriſtopher Nevile, Eſq. 2 L. P.
Mr. Nevile, M. A. late Fell. of Jef. Coll. Ca. L.P.
Sir Roger Newdigate, Bart. L. P.
Mr. Newnham, M. A. of C. C. C. Oxf. L. P.
Rev. Mr. J. Newte, Fell. of C. C. C. Oxf. L. P.
John Nicolſon, Eſq. of Carliſle.
Rev. Dr. Nicolſon, Fell. of Queen's Coll. Ox. L.P.
Anth. Norris, Eſq. of Barton, Norfolk. L. P.
Rev. Mr. Norris. L. P.
Rev. Dr. Nowell, Principal of St. M. Hall. Ox. L.P.
Rev. Cradock Nowell, St. Mary-hall, Oxf. L. P.
Nicholas Nugent, Eſq. L. P.

O.

Thomas Okes, M. D. of Exeter. L. P.
Mr. Oliver, Fell. Com. of Clare H. Camb. L. P.
Paul Orchard, Eſq. Col. of N. Dev. Militia. L. P.
Dr. Oſborn. L. P.
Mrs. Oſborn. L. P.

Rev. Dr. Owen, Rector of St. Olaves, Hart-Str,
Mr. Owen, of St. John's Coll. Oxf.

P.

Rev. Mr. Paley, Prebendary of Carlifle.
John Palmer, Efq.
John Paradife, Efq.
Rev. Mr. Parkinfon, M. A. Fell. of Chrift's C. Ca,
Rev. Mr. Parminter.
Rev. Dr. Parr, Mafter of Norwich-School. L. P.
Mr. Parr, of C. C. C. Oxf. L. P.
John Parry, Efq. M. P. L. P.
John Parfons, M. D. of Chr. Ch. Oxf. 2 L. P.
Mr. Parfons, of Wadham Coll. Oxf. L. P.
Mr. Partington, of Chr. Ch. Oxf. L. P.
John Peachey, Efq. L. P.
Rev. W. Pearce, B. D. Publ. Orat. of the Univ. C.
Rev. Mr. Peele, Norwich. L. P.
Granville Penn, Efq. of Wadham Coll. Ox. L. P.
Dr. Pennington, of St. John's Coll. Camb. L. P.
Lucas Pepys, M. D. L. P.
Rev. Dr. Percy, Dean of Carlifle. L. P.
Mr. T. Percy, of Emanuel Coll. Camb. L. P.
Edward Peter, Efq. of Trinity Coll. Oxf.
The Lord Bifhop of Peterborough. L. P.
Rev. Mr. Pett, Stud. of Chr. Ch. Oxf. L. P.
William Pitcairn, M. D. Pref. R. Coll. Phyf. 4 L. P.
Mr. Pitfield, of Exeter. 2 L. P.

Rev.

Rev. J. Plumptre, M. A. Vicar of Stone.

Edward Poore, Efq. L. P.

Rev. Mr. Popple, M. A. Fell. of Trin. Coll. Camb.

Rev. Mr. Porter, M. A. Fell. of Trin. Coll. Camb.

The Dutchefs Dowager of Portland. 2 L. P.

Rev. Mr. Poftlethwaite, M. A. Fell. of Tr. C. L.P.

W. M. Praed, jun. Efq. L. P.

Rev. Mr. Prefton, M. A. of Trin. Coll. Camb.

Major Price, of the King's own Dragoons.

Rev. R. Price, LL. B. Fell. of A. S. C. L. P.

Rev. Mr. Pritchett, Fell. of St. John's Coll. Camb.

Robert Prudom, Efq. of Exeter.

Mr. Putt, of C. C. C. Oxf. L. P.

Q.

Rev. Nut. Quicke, Chan. of the Ch. of Ex. 2 L.P.

Andrew Quicke, Efq. of Ethy, Cornwall. L. P.

R.

Mr. Radcliffe, of Trinity Coll. Camb.

Rev. Dr. Randolph, Prefident of C.C.C. Ox. L.P.

Francis Randolph, Efq. L. P.

Rev. Mr. Randolph, Stud. of Chr. Ch. Ox. 2 L.P.

Rev. Mr. Randolph, of Chr. Ch. Oxf. L. P.

Rev. Mr. Randolph, of Magd. Coll. Oxf.

Mr. Raftall, M. A. Fell. of Jefus Coll. Camb.

Rev. Mr. Ratcliffe, Fell. of Brafennofe, Coll. L. P.

Rev. Mr. Rawes, Fell. of Queen's Coll. Oxf.

Mrs,

Rev. Mr. Raynes, M. A. Fell. of Jefus Coll. Camb.

Dr. Remmett, Phyfician at Plymouth.　L. P.

The Rev. Mr. Rennell.　2 L. P.

Mrs. Reynolds.　L. P.

Sir Joſhua Reynolds.　L. P.

H. R. Reynolds, M. D.　L. P.

Ambrofe Rhodes, Efq. of Bellair, Devon.　L. P.

Rev. Mr. Rhudde, Vicar of Shepherd's well, 2 L.P.

Rev. C. Richards, B. A. of C. C. C. Oxf.　L. P.

Rev. Dr. Robert Richardfon.　L. P.

Rev. Mr. Richardfon, Minif. of St. Cuthbert's, Carl.

Rev. H. J. Rickman, of C. C. C. Oxf.　L. P.

Right Hon. Lord Rivers.　2 L. P.

Mr. Roch, of Barnſtaple.

Rev. Mr. Rogers, Stud. of Chr. Ch. Oxf.　L. P.

Giles Rooke, Efq. Serjeant at Law.　L. P.

Mr. Rookes, Fell. Comm. of Jefus Coll. Camb.

Walter Ruding, M. D. Fell. of Merton Coll.　L. P.

S.

Theoph. R. Salway, Efq. B.A. Fell. of A. S. C. L.P.

Rev. Dr. Sandby, Chanc. of Norwich.　L. P.

The Lord Sandys.　2 L. P.

Mr. William Sanford, Exeter.　L. P.

Mr. George Savage, M.A. Fell. of King's C. Camb.

Sir George Savile, Bart. 6 L. P.

Mr. Saunders, B. A. Fell. of Queen's Coll. Camb.

Dr. Saunders.　L. P.

Rev. Mr. Schomberg, of Magd. Coll. Oxf.　L. P.

d

Alex-

Alexander Scot, Efq. L. P.

Rev. Mr. Seger, of C. C. C. Oxf. L. P.

Rev. W. Segrave, B. D. of Trinity Coll. Oxf.

—— Seward, Efq. 2 L. P.

Rev. Mr. Shepherd. 2 L. P.

R. B. Sheridan, Efq. L. P.

Mr. William Sharp. 20 L. P.

Humphrey Sibthorpe, Efq. L. P.

J. Sibthorpe, Efq. M. A. Ratcl. Fell. U. C. Ox. L.P.

Rev. Ja. Simons, Rect. of St. Stephens, Exeter.

Rev. Jo. Simons, of Heavitree, Devon. L. P.

Rev. Mr. Siſſon, Stud. of Chr. Ch. Oxf. 2 L.

Rev. Mr. Skinner, of Baſſingham. L. P.

The Lord Chief Baron Skynner. 2 L. P.

Rev. Dr. Skynner, Præcentor of Exeter. L. P.

Rev. Mr. Sleech, Archdeacon of Cornwall. L. P.

Mr. Sloley, Caius Coll. Camb.

Rev. Dr. Smallwell, Canon of Chr. Ch. Oxf. 2L.P.

John Smith, M. D. Sav. Prof. Geom. Oxf. L. P.

Dr. Hugh Smith, Hatton-ſtreet.

Rev. Mr. Snell.

John Speare, Efq. of Exeter. L. P.

Earl Spencer. L. P.

Counteſs Spencer. L. P.

Rev. Leigh Spencer, M. A. Fell. of A. S. Coll.

Rev. Mr. John Spurway. 4 L. P.

Rev. Mr. William Spurway.

Rev. Mr. Squire.

George Steevens, Efq. 10 L. P.

John

John Stephens, Efq. of Coaver, Devon. L. P.

Mr. Stephens.

Rev. Mr. Stephenfon, M. A. Fell. of C. H. Ca. L. P.

Rev. Mr. Stevens.

Rev. Dr. Stinton, Chanc. of Lincoln. 20 L. P.

Rev. Mr. Stinton, Fell. of Exeter C. Oxf. 10 L. P.

Richard Stonehewer, Efq. L. P.

Rev. Mr. Storer, Norwich. L. P.

The Lord Vifcount Stormont. L. P.

Rev. Humphrey Sumner, M. A. 2 L. P.

Rev. Mr. Sweet.

Dr. Symonds, Prof. Mod. Hift. Camb. L. P.

T.

Rev. Mr. Tapps, Norwich. L. P.

Rev. Dr. Tarrant, Dean of Peterborough. 2 L. P.

Edward Taylor, Efq. L. P.

John Taylor, M. D.

Rev. Mr. Templeman, Rect. of Shafton St. Ja. Dor.

Sir Noah Thomas, M. D. Phyfician to the K. 2 L P.

Rev. Mr. Thomas, Lect. of St. Olaves, Hart-ftreet.

Sir John Thorold, Bart. L. P.

Mr. Todd, Bookfeller, at York.

Colonel Townfend.

Hon. and Rev. J. Tracy, D. D. Ward. of A.S.C. L.P.

Rev. Mr. Trimenhere, of Trin. Coll. Camb.

Fiennes Trotman, Efq. 2 L. P.

Rev. Mr. Tutté, Stud. of Chr. Ch. Oxf. L. P.

Rev. Mr. Turner, M. A. Fell. of P. H. Camb. L. P.

Richard

Richard Turton, M. D. Phyf. Ext. to the K. L. P.
Mr. R. Cooke Tylden, of Jefus Coll. Camb.
Mr. Tyrwhitt.
Rev. Mr. Tyrwhitt, of Wickham, Effex. L. P.
Rev. Mr. R. Tyrwhitt, M. A. Jefus C. Cam. L.P.
Mr. T. Tyrwhitt, Stud. of Chr. Ch. Oxf. L. P.
Richard Tyfon, M. D. L. P.

V.

Rob. Vanfittart, Efq. LL. D. Fell. of A.S.C. 2 S.P.
William Vivian, M. D. Reg. Prof. Med. Oxf. L. P.
Rev. Dr. Vyfe, F. R. S. and A. S. L. P.

W.

Rev. Mr. Wall, M. A. Fell. of Chrift's Coll. Camb.
Rev. Mr. Wall. L. P.
Mr. Warburton of Jefus Coll. Camb.
Dr. Waring, Lucas. Prof. of Math. Camb.
Mr. Warre, of Jefus Coll. Camb.
Richard Warren, M. D. Med. Reg. L. P.
Rev. Dr. Warton, Mafter of Winchefter-fchool.
Rev. Dr. Watfon, Reg. Prof. of Divinity, Cam. L.P.
Sir Charles Watfon, Bart. M. A. F. of A. S. C. L.P.
Rev. Benjamin Webb. L. P.
William Webber, Efq. 2 L. P.
Hon. Tho. Wenman, LL. D. Fell. of A. S. C. L.P.
Earl of Weftmorcland. L. P.
Rev. Mr. Wefton, Prebendary of Durham. L. P.
Rev. Stephen Wefton, Rector of Mamhcad, Devon.

Rev.

Rev. Mr. Weſton. L. P.

Rev. Dr. Wheeler, Canon of Chr. Ch. Oxf. 2 L.P.

Mr. Whicher, C. C. C. Oxf. L. P.

Samuel Whitbread, Eſq. of Chr. Ch. Oxf. L. P.

James White, Eſq. of Exeter. 2 L. P.

Rev. Mr. Whitchurſt, of Ickleford, Hertfordſhire.

Mr. Whitehurſt, of Catharine H. Camb. L. P.

Mr. Whitehurſt, of Peterhouſe, Camb. L.P.

Rev. Jo. Whitfeld, Rect. of Bideford, Devon.

Mr. W. Wickham, Stud. of Chr. Ch. Oxf. L. P.

Ralph Willet, Eſq. L. P.

Mr. Williams, of C. C. C. Oxf. L. P.

Mr. W. Wilſhire, jun. of Hitchin, Hertfordſhire.

Rev. Dr. Wilſon, Canon Reſidentiary of St. Pauls.
2 L. P.

Rev. Mr. Wilſon, M. A. Fell. of Chriſt's C. Camb.

Mr. Benjamin Wilſon. 2 L P.

Rev. Mr. Wilſon, Fell. of Queen's Coll. Oxf.

The (late) Lord Biſhop of Wincheſter. 2 L. P.

John Withers, Eſq.

Rev. Mr. Wodehouſe, Prebend. of Norwich. 2L.P.

Michael Wodhull, Eſq. L. P.

Rev. Mr. Wood, Fell. of Cath. Hall, Camb.

William Woodeſon, Eſq. Viner. Prof. Oxf.

Mrs. Woodifield. 2 L. P.

Rev. Mr. Worth. L. P.

Daniel Wray, Eſq.

Sir Cecil Wray, Bart. 2 L. P.

William Wright, Eſq. L. P.

J. E. F. Wright, Eſq. L. P.

Richard

Richard Wright, M. D. · L. P.
Mr. Wyatt, M. A. Fell. of Pemb. H. Camb. L. P.
Jofeph Wyndham, Efq. L. P.
Rev. Luttrell Wynne, LL. D. Fell. of A. S. C. L.P.

Y.

Rev. Dr. Yates, Maft. of Cath. Hall. Camb. L. P.
Edward Roe Yeo, Efq. L. P.
Rev. Mr. Duke Yonge. L. P.
The Lord Archbifhop of York. 2 L. P.
Mr. C. Young, Surgeon at Plymouth. L. P.

Right Hon. Lord John Cavendifh. L. P.
Sir John Duntze, Bart. of Rockbere, Dev. 2 L. P.
Michael Morris, M. D. 2 L. P.

DISSERTATION I.

ꭍ

ON THE

GRÆCIAN MYTHOLOGY.

DISSERTATION I.

ON THE

GRÆCIAN MYTHOLOGY.

THE affertion of HERODO-
TUS *, " that the Theo-
" logy of the Greeks was no older
" than the times of HOMER and
" HESIOD," is, I think, fully
overthrown by PAUSANIAS, whofe
refearches into the antiquities of
his country were much more
profound and exact, than thofe
of any other writer that is come
down to us. HERODOTUS in particular, who had travelled into

* L. II. c. 53.

B   feveseveral neighbouring countries, to collect materials for those parts of his history, appears not to have been equally attentive to the ancient traditions of Greece. Of the kings of Lacedæmon he has given a very erroneous list *; and in respect to the age and family of Lycurgus † is contradicted by every other writer. Nor are there any traces of his having perused several ancient Poets and Genealogies, which appear to have been extant even in the time of Pausanias. This latter quotes a verse from Pamphos‡, a writer of Hymns, which mentions the two different attri-

* Herod. VIII. 131.
† Herod. I. 65.
‡ Pausan. p. 577. Ed. Kuhn.

butes

butes of NEPTUNE, his being
the author of Chariots and of
Ships. And PAMPHOS he affirms
to have lived not only before
HOMER, but before the Trojan
war, and even before NARCIS-
SUS *, whom OVID makes contemporary with TIRESIAS †.

Besides, the very perusal of
Homer shews, in my opinion,
that he was not the author of
his Mythology; as he never
attempts to explain it, but sup-
poses his reader, or rather his
hearer, sufficiently acquainted
with it. To which add, that
several of the Græcian temples

* Paufan. p. 773. Ed. Kuhn.
† Metam. III. 340. feq.

were

were in being long before Homer, particularly that of Diana at Aulis in Bocotia, which
Pliny informs us * was *seculis
aliquot ante Trojanum bellum ex-
ædificata.*

There feems to be juft as little reafon for the affertion of
others, that a part of Greece, and
particularly Athens, was peopled
by Colonies from Ægypt. This
opinion is countenanced by Strabo † and Diodorus ‡ Siculus ;
but the firft broacher of it, I believe, was the Hiftorian Theopompus. So fays Proclus § ; and

---

* L. XVI. c. 40.
† Strabo, L. VII. p. 321.
‡ Diod. Lib. I. c. xxviii. 9.
§ In Platon. Timæum, p. 30.

also,

alſo, that he was flatly contra-
diſted by others, who charged
him with ſpreading this ſtory
out of mere prejudice. The
Athenians alſo, as we learn from
LUCIAN *, conſidered it as the
height of paradox to talk of CE-
CROPS as a foreigner. Let us
conſider the authorities therefore
as equal, and weigh the ſtory
according to probability. We
know a good deal of the Ægyptian cuſtoms, though not much,
with certainty, of their hiſtory.
The Græcian cuſtoms we know
ſtill better. Now between theſe
two there is a total diverſity.
The Ægyptians were circumci-
ciſed : the Greeks held that prac-

* Περὶ ἀποῤῥάδος. II

B 3   tice

tice in contempt and derifion. The Ægyptians indulged themfelves in a plurality of wives. The Greeks were permitted to have no more than one; and of this law Cecrops, the pretended Ægyptian, was the author. In Ægypt, according to Sophocles*, weaving and other domeftick bufinefs was carried on by the men ; and the care of providing food was left to the Female. In Greece the contrary and more rational cuftom prevailed. In Ægypt it is faid to have been a rule, that the Son fhould follow his father's profeffion : In Greece no traces of fuch a rule are to

---

* Oed. Colon. 330. Philochorus, ap. Scholiaft. ibid.

be

be found. The Ægyptians worfhiped Animals and Plants ; the
Greeks defpifed and ridiculed
this fuperftition. The Ægyptians
paid a fcrupulous attention to
nativities and the afpects of the
Planets : In the Græcian Hiftory,
among the various ways of en-
quiring into futurity by oracles,
the flight of birds, infpection of
entrails, and the like, there is
not a fingle inftance of any at-
tempt to calculate nativities.
Laftly, the Ægyptians were particularly ftudious to preferve the
dead body from diffolution by
their careful and coftly method
of embalming; whereas the
Greeks, by committing it immediately to the Fire, feem to have

been defirous of promoting its diffolution.

Now, if Greece was not peopled by colonies from Ægypt, according to the affertion of THEOPOMPUS, we have no authority for rejecting on that pretence the traditions of the Greeks with refpect to their own origin. And thefe inform us, that the bulk of the nation were indigenous; the offspring, if I may fo fpeak, of the foil; and that the firft foreign fettlers among them were DANAUS, PELOPS, and CADMUS.

After this it is perhaps needlefs to refute another affertion of

HE-

HERODO₁us * " that the Greeks
" received the names of their
" Gods from the Ægyptians."
For if he meant to fay, that Ho-
MER borrowed from the Ægyptians
that Theology, or lift of Gods,
which he publifhed in Greece,
this has already been˙refuted
upon the beſt authority. And
if he meant that the Greeks received that lift from the Ægyptians at ſo nₑ earlier period, then
he contradiᑕs himſelf, and de-
ſervₛ nₒ crₑdic either for the one
or the other.

Eu, to put HERODOTUS for the
preſent ouꞇ of the queſtion,
what foundat n is there for ſay-

* Hₑrod. II. 4.

ing

ing that the Greeks received the names of their Gods from the Ægyptians? Is THOTH the fame name as HERMES *, or ARSAPHES as DIONYSUS †, or ARUERIN as APOLLON ‡, or NEPHTHEN as APH-RODITE §, or NEITH as ATHENE ‖? There were indeed fuch cities in Ægypt as DIOSPOLIS, HELIOPO-LIS, and PANOPOLIS; but were thefe names ufed by the Ægyp-tians, or *only* by the Greeks? Moft certainly the latter; becaufe I find, in THEOPHILUS of Anti-och, that the Ægyptian word

---

* Clem. Alex. Strom. I. p. 303. Ed. Par.

† Plut. de. Ifid. & Ofir. p. 365.   Ed. Frankf.

‡ Plut. ibid. p. 355, 6.

§ Plut. ibid.

‖ Plato in Timæo, p. 21. 1043.   Ed. Ficin.

for

for HELIOPOLIS was ON *; and in DIODORUS SICULUS, that PANOPOLIS was called by the Natives CHEMMO †.

But if not the names, perhaps they derived the divifion and attributes of their Gods from Ægypt. This is a thing much more difficult of proof. It feems not impoffible, that feveral nations, not communicating with one another, might have each of them a fupreme prefiding God; a God of War; a God of Love; a God of Love; a God of Eloquence, and fo forth. Afterwards, when an intercourfe is

* Theophil. ad Autol. III.
† Diod. Sic. Lib. I.

begun.

begun, they each expect to find their own Gods worshiped under some denomination, or other, by the stranger nation, and therefore readily find out the points of resemblance. Thus, when the Latins and Greeks began to converse familiarly together, they soon discovered a general resemblance between ATHENE and MINERVA; between APHRODITE and VENUS; between ARTEMIS and DIANA; although the attributes of the Latin Deities (such attributes, I mean, as were assigned them by popular superstition) are far from corresponding exactly to those of the Græcian. However, there was similitude enough to encourage the Latin

Poets

Poets to apply to their own Deities every thing, without referve, that had been faid of the Græcian. And hence in procefs of time they came to be confidered as identically the fame. But that the Romans did not borrow their Gods from the Greeks appears from this circumftance, that many of the Græcian Deities * have no correfpondent Latin Deities ; and many of thefe latter † have no archetype among the Gods of Greece.

In confirmation of this we may obferve, that the Greeks were fagacious enough to find

* Latona, Priapus, Hecate.
† Janus, Vertumnus, Flora, Pomona.

out

out their own Gods in Perfia, notwithftanding the known contrariety of the two religions. The Perfians, we are told *, worfhiped JUPITER under the name of OROMASDES; HADES, under the name of ARIMANIUS †; VENUS, under the name of MITRA; and DIANA, under the name of ANITIN ‡. It was the fame thing with refpect to the Romans and the Gauls. The hierarchy, and the many peculiar fuperftitions of the Druids, are fo totally foreign to every thing we know of the Romans, as to pre-

* Ariftot. apud Diog. Laert. in Procemio.
† Herodot. I. 131.
‡ Plut. in Artaxerxe, p. 306. Agathias, II. p. 45. Ed. Venet.

clude

clude every idea of one nation having borrowed from the other. Yet Cæsar does not hefitate to fay *, that the Gauls worfhiped Apollo, Mars, Jupiter, and Minerva ; meaning, I fuppofe, certain Gods correfponding in their attributes to thofe fo denominated by his countrymen. The Gods of Scandinavia in later times were confubftantiated in like manner with the Gods of Latium.   No one, I imagine, will pretend that their Thor, and Woden, and Friga, were the copies of Jupiter, Mercury, and Venus.   And yet when the Goths came to intermix with the Romans, they found refem-

* De Bell. Gall. VI. 15.

blance

blance enough between their Gods to tranflate one of thofe words by the other : which cuf-tom appears to have been fo uni-verfal, that we have no other way of rendering WEDNESDAY, THURSDAY, and FRIDAY into Latin, but by calling them DIES MERCURII, JOVIS, and VENERIS.

Such an imperfect refemblance as this might poffibly obtain be-tween fome of the Ægyptian Gods and fome of the Græcian. But that this refemblance was general, or in any cafe fo exact as to indicate imitation, I fee no reafon to believe. What traces have we among the Greeks of the worfhip of CROCODILES, and

and SERPENTS? of the Bull APIS, or the Dog ANUBIS? Has not VIRGIL * exprefsly fet the Ægyptian Gods in oppofition to the Græcian? and would he have done this, if the former had been archetypes of the latter, or even if the oppofition between them had not been ftriking?

Upon the whole, therefore, we have the beft reafon to conclude, firft, that the Greeks in general were an indigenous people, αὐτόχθονες: and, fecondly, that their RELIGION and MYTHOLOGY was radically, if not entirely, their own. And now having cleared the ground, I

* Æneid. VIII. 698.

C   fhall

fhall endeavour to place that MYTHOLOGY upon its proper foundation.

This intricate fubject will, I think, be better underftood, if we divide the GRÆCIAN MYTHO-LOGY into two claffes; that which is effential, and that which is acceffory. The effential I would call the worfhip of the SUPERIOR GODS, fuch as SATURN, JUPI-TER, NEPTUNE, PLUTO, APOL-LO, MARS, MERCURY, VULCAN, BACCHUS, JUNO, PALLAS, CERES, DIANA, VENUS, CYBELE. Thefe may be confidered as fo many allegorical perfonages, reprefent-ing either the great divifions of Nature, as the Heaven or up-

I   per

per fky, the Air, the Sea, the Earth, the fubterraneous world; or elfe thofe operations and qualities, which have a more particular influence upon the Animal world and upon Society. Of this latter kind are the propagation of Animals, Tillage, Handicrafts, War, the art of Mufic, Divination, Hunting, and the Palæftra. The particular divinities prefiding over each of thefe departments need not be pointed out to the claffical reader. The only one liable to be miftaken is that of JUNO, who has been thought by fome to reprefent the earth: a falfe notion, founded, I apprehend,

C 2   upon

upon thefe verfes of VIRGIL *:

Tum Pater omnipotens fœcundis imbribus æther

Conjugis in gremium lætæ de- fcendit——

Whereas VIRGIL feems here to have had neither JUPITER nor JUNO in his eye, but to have alluded to the more ancient fable of Ουρανος and Γη, as expreffed in a fragment of the OEDIPUS of EURIPIDES:

Ἐρᾷ δ᾽ ὁ σεμνὸς ὀρανὸς, πληρέμενος
Ὄμϐρϐ πεσεῖν εἰς γαῖαν Ἀφροδίτης ὑπο.

which LUCRETIUS † has alfo a- dopted :

Poftremo pereunt imbres, ubi eos PATER ÆTHER

* Georg. II. 325.
† Lib. I. 251.

In

In gremium MATRIS TERRAI
præcipitavit.

According to PHILO JUDÆ-
US *, JUNO is not the Goddefs
of the Earth, but of the Air:
Ἥραν δὲ τὸν ἀέρα, ἢ τὸ πῦρ Ἥφαι-
ςον, ἢ ἥλιον Ἀπόλλωνα—μυθογράφοις
παρέδοσαν. Even the more mi-
nute parts had their appropriated
Gods, though of inferior rank
and power, in proportion as the
fubftance to which they were
annexed was of a greater or lefs
confequence in the vifible world.
Rivers and brooks from the per-
petuity of their ftream naturally
excite wonder; and that wonder
foon begets an idea of divinity.

* Philo Jud. p. 513. Ed. Turn. See alfo
Cic. de Nat. Deor. II. c. 26.

C 3   But

But the River from its fuperior magnitude being a more awful object was put under the tutelage of a more important and mafculine God ; whereas the Brook, which fuggefted only pleafurable ideas without any mixture of terror, was fuppofed to derive its origin from a tender Female. Nymphs, that is, Goddeffes of inferior rank, were in like manner confidered as inhabiting and protecting Trees, whofe apparent life was naturally enough attributed to the power of an inherent Deity.

Thefe I confider as the effential parts or *flamina* of the GRÆCIAN MYTHOLOGY. The accessory acceffory confift of either fome wonderful *phænomena* of Nature, or of fome extraordinary Hiftorical Facts, told in an allegorical manner, and improved into miracles. I fhall give fome clear and undoubted inftances of both forts, which will furnifh the reader with a clue to interpret the reft.

Among natural *phænomena* there is hardly any more awful than that of volcanoes, or burning mountains, in their ordinary ftate; but in a time of violent eruption they become tremendous. It may be fufpected that the ancient fable of the Giants attacking the Gods was only an

C 4.   allegorical

allegorical picture of thofe eruptions, which by the great quantities
of melted minerals and afhes
which they throw up, feem in
effect to be making war againft
heaven. This notion is ftrongly
countenanced by Strabo *; who
informs us, that the part of Cilicia,
where Typhon was fuppofed to
refide, is called *Katakekaumene*,
or the Burnt Country, from the
cineritious appearance of the
earth. The commotions and
occafional eruptions of Ætna are
attributed, we know, to Typhon
and Briareus in another way.
Thofe *phænomena* not ceafing
even in the later ages, the Poets
were no longer able to impute

* L. XII. p. 579.

them

them to the war of the Giants
againſt JUPITER, becauſe that
would have implied that the
Giants were ſtill unconquered,
and in a condition to renew
the attack. They therefore give
the ſtory another turn ; and tell
us, that one or other of thoſe Giants is buried under Ætna, and
produces a freſh commotion of
that mountain as often as he
changes his poſture of lying on
his right or left ſhoulder. The
Solfatara, or Burning Country,
near Naples, is accounted for by
the giants being buried there *,
and the thunder yet unquenched
ſending up a vapour through the
crevices of the earth : though

* Strabo, l. V. p. 248.

others

others indeed go fo far as to fay, that this was the true fcene of the giants' war. Now, as we find that volcanoes, and countries abounding with natural fire, were attributed one way or other to giants, it is no extravagant conjecture to fuppofe, that the very exiftence of giants was originally built upon thefe *phænomena*.

This however is not the only fable, to which volcanoes have given rife. In one or two inftances, where the difcharge of flame has been moderate and equable, the cavity of the mountain has been confidered by the Poets as the workfhop of VULCAN. The ifland of Lipara is one

of

of thefe fhops; and Mofyclus, a hill in the Ifland of Lemnos, another. The latter is thus defcribed by VALERIUS FLACCUS * :

Ventum erat ad rupem, cujus pen-
  dentia nigris
Fumant faxa jugis, coquiturque
  vaporibus aer :
Subftitit Æfonides : atque hic Re-
  gina precari
Hortatur, cauffafque docens, hæc
  antra videtis,
Vulcanique, ait, ecce domos.——

It is well known that feveral fmall Iflands have been protruded from the bottom of the Sea by volcanoes, which protrufion has in fome cafes been effected gra-

* Lib. II. v. 336.

dually,

dually, and in others by a single
eruption.   If we suppose the
Island of Delos to have continued a considerable time about
the level of the sea, it would of
course be sometimes visible to
the mariner, and sometimes in-
visible, according to the calm-
ness or roughness of the water.
It is possible also that the top
of the Island, after being some
time above the surface, might
sink down below it for want of
support, as the new-formed hills
about Vesuvius not unfrequently
do.   In either way the appearance of the island at one time,
and its disappearance at another,
sufficiently accounts for the fable of its swimming from place

to

to place. That this Ifland
was formerly under water, and
difcovered by the gradual wafh-
ing of the fea, is affirmed by PHI-
LO *, upon the authority of
ancient Hiftories. This fufficiently juftifies my interpretation
of the fable ; though its final
appearance is, I think, better
accounted for from the opera-
tion of a volcano, efpecially as
it was not a flat Ifland, but had
a confiderable eminence upon it,
called Cynthos.

Earthquakes were common in
Greece, and were attributed in
their Mythology to one general

* Περι αιδιος. κοσμ. p. 658. Ed. Tur-
neb.

caufe,

caufe, the elevation of the earth by the trident of NEPTUNE. There is however one fable which appears to have a direct reference to them, though not explained in that manner by Mythologifts. TANTALUS, the King of Phry-gia, and father of PELOPS, re-fided in a city called Sipylus, fituated upon a mountain of the fame name *. This City was either entirely thrown down, or greatly damaged, by an earth-quake during the reign of TAN-TALUS; who, after that, we may fuppofe, lived in continual dread of the like event. The punifhment therefore which the Poets contrived for him after

* Strabo, L. I. p. 58. L. XII. p. 579.

Death,

Death, that he was every moment in dread of being crushed by a stone that hung over him, is nothing more than the situation in which every man must live, who has experienced, like TANTALUS, the horrors of an earthquake.

The story of DEUCALION's Deluge I do not consider as a fable; and I likewise think it had no reference to the deluge of NOAH. It cannot be the latter, because the time of it is specified in the Græcian Chronology *, according to which there must have been an interval of at

* Clem. Alexand. Strom. I. p. 321. Ed. Per.

leaft

leaſt a thouſand years between the two. The tract of country affected by it is particularly mentioned by ARISTOTLE *. It did not even extend to the Peloponneſus, but was in a great meaſure confined to the plains of Theſſaly. From theſe circumſtances it appears to have been a very diſtinct event from the general deluge; and the ſame circumſtances furniſh alſo a ſtrong preſumption that it was not entirely fabulous.

The burning of the world by PHAETHON muſt be in part hiſtorical, becauſe the Græcian Chronologers, according to CLE-

* Meteorolog. I. p. 32. Ed. Sylburg.

MENS

MENS * fixed the time of it to a year. The event, that gave rise to this fable, is univerfally allowed by Mythologifts to have been a long continuance of heat and drought. And the reality of fuch an event, which is in itfelf fufficiently probable, receives fome confirmation from two or three fcraps of Hiftory that are come down to us. CALLIMACHUS † fays that in former times there was a drought in Ægypt for nine years:

Αἴγυπ]ος προπάροιθεν ἐπ᾽ ἐννέα κάρφε]ο ποιάς.

* Strom. I. p. 332. 335. Ed Par.
† Fragm. CLXXXII. See Hyginus, Fab. LVI.

D   And

And a fimilar tradition is fome-
where mentioned by PAUSANIAS,
as being preferved in one of the
obfcure towns of Greece.   But
whence arofe the fable of PHAE-
THON driving the chariot of the
Sun ? It might be fuggefted by
the derangement obferved in the
common courfe of nature, and
by the improbability, that the
Sun, whofe ordinary influence
is   fo   beneficial   to   mankind,
fhould, under the management
of the fame beneficent Gover-
nor, become ruinous and deftruc-
tive.   From this feeming contradiction the Mythologift eafily
extricated himfelf by the fiction
of a new and unexperienced
Charioteer.   Or if we fuppofe
with

with PROCLUS *, that this extraordinary drought was occafioned
by a Comet, the difappearance
or extinction of which, accor-
to PORPHYRY †, was commonly
called κερκύνωσις, this would fhew
us, why the Poets defcribe their
PHAETHON as taking fuch an
excentrical courfe, and finally
deftroyed by lightning.

The grotefque figures of rocks
furnifhed alfo fome embellifhment to the GRÆCIAN MYTHO-
LOGY. To fay nothing of thofe
in the *Fretum Siculum*, which
fuggefted the fable of SCYLLA
and her Dogs; there was a re-

* in Timæum, p. 33, 34.
† Ibid. p. 34.

D 2 markable

markable one in Mount Sipylus, which at a distance presented the lineaments of a Woman in great dejection and drowned in Tears, though upon a nearer view, according to Pausanias *, the resemblance was lost. The Mythologist found no difficulty in naming the person by whose Metamorphosis this rock was produced. Niobe, the Daughter of Tantalus, was a native of this very spot; and the severe affliction, which she must have felt from the loss of her numerous progeny, naturally pointed her out as the original cause of this peculiar appearance.

* Attic. p. 48.

Philos-

PHILOSTRATUS informs us *, that, near the town of Phæſtus in Crete, there ran out into the ſea a promontory, terminating in a rock, which had the exact form of a Lion. It was natural that ſuch a rock ſhould be ſuppoſed to have had ſome diſtinguiſhed origin: and accordingly the Inhabitants reported, that this was one of the Lions that anciently drew the car of the Goddeſs CYBELE. The Boeotians equally ingenious, having in their country two naked rocks, reſembling, I ſuppoſe, two animals, pretended that one was the Fox of Teumeſſus, whoſe deſtiny it was never to be caught; the

* Vit. Apollon. L. IV. c. 34.

D 3   other

other the Dog of CEPHALUS,
whose destiny it was, that nothing should escape from him :
which two contrary destinies
JUPITER, according to them, re-
conciled by turning them both
into stone *.

There is in Barbary, according
to Dr. SHAW, a large plain, or
tract of country, scattered over
with great numbers of naked
rocks, standing erect, and so
proportioned their thickness to
their height as to carry the appearance of Men. He conjectures, and, I think, with great
probability, that this assemblage
of natural statues suggested the

* Antonin. Lib. c. 41.

idea

idea of the Inhabitants of Africa having been turned into ſtone by Perseus, when he invaded and plundered the maritime part of that country. Seriphus, a little Iſland near Argos, where Perseus is ſaid to have performed another of thoſe miracles, was equally remarkable for numbers of naked rocks.

There was in the Iſland of Crete a Plane-tree, remarkable for not ſhedding its leaves in Winter *. Such a ſtriking exception to the common rule implied of courſe the action of ſome divinity : and the reaſon aſſigned by the Vulgar was,

* Theophraſt. Hiſt. Plant. Lib. I. cap. 15.

that under this tree JUPITER firſt obtained poſſeſſion of the fair EUROPA. This Plane-tree, according to THEOPHRASTUS, grew near a fountain; with reſpect to which ANTIGONUS CARYSTIUS aſſerts *, that thoſe who ſat round it in time of rain were not liable to be wet. A fact which none will believe, requires no explanation. Yet it is worth while to remark the progreſs of credulity. A natural ſingularity created firſt a religious veneration in the beholder, and that in its turn produced a fictitious miracle.

* Antigon. Caryſt. cap. 179.

The

The principles of Chemiſtry were unknown to the ancients; and therefore, when they met with any mineral ſpring, inſtead of analyſing it to find out the ingredients from which it derived its properties, they were con-tented to derive them from the interpoſition of ſome Deity, or from ſome remarkable event of the Mythologic Age. A ſalt ſpring in Illyria, according to the popular report preſerved by ARISTOTLE, was an act of boun-ty from HERCULES to the in-habitants of that country. A river in Elis, whoſe water was fœtid, was ſaid to have contract-ed that fœtor from the arrows of HERCULES being waſhed in

it,

it, after having been fmeared
with the gall of the Hydra:
though about this the orthodox
of thofe days were not agreed,
as fome imputed it to the ablu-
tions ufed to recover the daughters of PROETUS from their infa-
nity *.

It is pleafant to obferve the
different ufe made of the fame
fact in different ages. A *non-
defcript* bird or plant is in our
days a treafure to the Naturalift,
who is happy enough to difcover it, and it ferves him as a
ftep to afcend into the temple
of Fame. The ancient Mythologift applied it to a different

* Paufan. Eliac. I. p. 387.

purpofe.

purpofe. He commonly found
out fome unfortunate prince or
princefs, who finking under the
weight of calamity had been
permitted by the Gods to affume this new fhape. Thus
PAUSANIAS informs us * that
the *Epops*, or *Upupa*, was not
obferved till after the cataftrophe of TEREUS; nor the Hya-
cinth difcovered in Salamis till
the death of AJAX. Novelties
of the fame kind may poffibly
have fuggefted many fimilar
ftories : the appearance of new
fpecies of birds being a thing
taken notice of, after the age

* Paufan. Attic. p. 40—86.

of

of fables was paſt, by Antigonus Carystius * and Pliny.

I come now to lay open the
other ſource of Mythology,
which conſiſts, as I ſaid, in the
perverſion of hiſtory by allegorical and miraculous accounts
of common facts. Of this claſs
one of the moſt common fables is the deriving the birth of
every eminent perſon from one
or other of the Gods. This
they were tempted to do for
various reaſons. In the firſt
place, as our knowledge of
Antiquity is limited, and every
genealogy muſt begin ſome-
where or other, the Genealo-

* Antigon. Caryſt. cap. 132.

gift,

gift, when arrived at that point, would be obliged to confefs his ignorance of what went before, did he not, by making a God the bafis of his lift, put a ftop to further enquiry. All the Greek genealogies originate in this manner, fome from a River, fome from NEPTUNE, but much the greater part from JUPITER: by which we are to underftand nothing more than that the Genealogift was come to the end of his line, and had no real anceftor to fubftitute.

A fecond caufe that greatly increafed the number of thefe fpurious Gods, was, I fuppofe, the convenient covering that
fuch

such a pretence gave to female frailty. The wife or daughter of an ancient Chieftain, who listened to the dictates of love, had this advantage above the modern Beauty, that she had a chance of deriving honour from her gallantries : and if her parents or husband concurred in the fraud, or even if they were unrelenting and sceptical, might still hope to be enrolled among the spouses of the Gods, provided her pretensions were not rendered abortive by an imprudent and unseasonable diffidence. To point out particular instances would be blasting that reputation, which the Ladies of Antiquity wished no doubt to pre-

serve

ferve to the lateft pofterity. But
it will be no breach of decorum
to obferve, that EURIPIDES. *
mentions it as a common practice
of indifcreet young women to
extenuate every amorous failing
by making fome God or other the
partaker of it.

The Gods had indeed a fet
of retainers, who fometimes
acted as deputies for them on
thofe occafions. The priefts and
fubordinate officers of their temples, who are reprefented as lead-
ing a life of feafting and drun-
kennefs (ἐν ἑορταῖς ἀεῖ καὶ μέθαις
ὄντες †) availed themfelves fome-

* Ion. v. 1513.
† Schol. in Hermog. p. 226.

times

times of their connexion with the Deity to obtain poſſeſſion of a young and beautiful vota-ry. JOSEPHUS * tells a remark-able ſtory of the prieſts of ISIS ſeducing a Roman Matron of great virtue, by perſuading her that the God ANUBIS was ena-moured of her, and ſolicited the honour of her embraces; in whoſe ſtead they introduced to her a young Roman, called DE-CIUS MUNDUS. And according to PAUSANIAS † the inhabitants of Temeſa in Italy were enjoined to build a temple to the *manes* of a man who had been ſtoned to death for a rape, and to of-

* Antiquitat. Lib. XVIII. cap. 3.
† Eliac. II. p. 467.

fer

fer to him every year the moft
beautiful of their Virgins, upon
whom the immortal ravifher appears to have exercifed both his
luft and his cruelty. This practice continued for a long fpace
of time, till EUTHYMUS, an
Olympian Boxer, happening to
pafs that way, became ena-
moured of the beautiful victim;
and undertook to be her champion, upon condition of becoming afterwards her husband.
For a perfon trained to athle-
tic combats, the inmate of the
temple, though a divinity, was
no match. He was conquered
by EUTHYMUS, and in defpair,
as the ftory informs us, threw himfelf into the fea, and was heard

of no more. Thefe well-attefted ftories render it probable, that it was not APOLLO, but his prieft, that follicited the embraces of CASSANDRA, and that the courtfhip of Io, fo particularly defcribed in ÆSCHYLUS *, was an artful contrivance, fimilar to that of the priefts of Isis in JOSEPHUS, except that the lover of Io was the Prieft himfelf. When any of thefe plots fucceeded, the innocent lady would of courfe afcribe her pregnancy to the God, by whofe name fhe had been feduced.

The fuppofed offspring of the Gods were ftill further mul-

* Prom. v. 645. feq.

tiplied

tiplied by a fourth caufe, the admiration which excellence of every kind naturally excites in the world. This was generally thought to denote, or at leaft it was conftrued as a mark of, divine parentage ; and the particular divinity picked out as the Father, was determined by the nature of the diftinction *. A good Archer was reputed to be the Son of APOLLO ; and I remember to have feen a paffage in fome ancient author, where the tradition is accounted for in this very manner. It was probably for fkill in divination that IAMUS, the Augur men-

* Euftath. in Iliad. A. p. 14.

tioned

tioned by PINDAR*, was suppofed to derive his birth from the fame God. THEAGENES, an Athlete of remarkable ftrength, was reported by his countrymen the Thafians to be the fon of HERCULES †. The great warrior and the crafty orator were in like manner celebrated, one as the Son of MARS, the other of MERCURY. Excellence in Horfemanfhip or in Navigation denoted the party to be the Son of NEPTUNE; and extenfive dominion, of JUPITER. DÆDALUS, the famous ftatuary, was celebrated as the fon of VULCAN,

* Ol. VI. v. 74.
† Paufan. Eliac. II. p. 477.

and

and APOLLONIUS * mentions one PALÆMONIUS, who attained the fame honour on account of his lamenefs. To call a fkilful and fortunate husbandman the fon of CERES would have interfered perhaps with a known fact; and therefore the mythologifts were obliged to compliment IA-SION in a different way, by pretending that CERES was fo enamoured of him as to admit him to her bed.

The meaning of thefe fables was probably as well underftood as a modern Poet is, who calls his Patron a fon of MARS; with this difference, that among the

* Apollon. Rhod. Lib. I. v. 202.

ancients

ancients the fact might not be entirely disbelieved. With respect to death, MYTHOLOGY makes no distinction of ranks. Yet in some deaths the Gods were supposed to be particularly concerned. It is observed by EUSTATHIUS *, that Homer attributes the sudden deaths of Men to APOLLO, as of Women to DIANA. It should seem however that this was not wholly confined to sudden deaths, as their arrows are said to have killed the children of AMPHION and NIOBE, who according to PAUSANIAS † died of the plague. CALLIMACHUS ‡ also ascribes the

* Euft. in Iliad. T.
+ Bœot. p. 721.
‡ H. in Dian. v. 126, 7.

deaths

deaths of women in childbed to
DIANA. Mr. MARKLAND on EU-
RIPIDES * remarks, that it was
cuftomary to attribute deaths
from any fudden cafualty to the
paffionate regard of fome God,
taking the beloved object from
the world to himfelf. Thus
ORITHYIA, blown, I fuppofe,
from a precipice by the wind,
was reported to be carried off
by BOREAS; HYLAS, drowned
in a fountain, to be pulled in
by the Water-nymphs: and AM-
PHIARAUS, who in the hurry
of his flight from Thebes fell
with his chariot into a deep
chafm, to have been fnatched
into the bowels of the Earth by

* Suppl. v. 929.

the

the Gods themfelves, as a reward of his virtue. This, as
appears from fome epigrams and
infcriptions, quoted by Mr.
DORVILLE *, was the common
confolation of every parent and
relation, to whom fuch difaf-
ters happened. That accidents
from lightning fhould be imputed
to JUPITER, is not to be wondered at, that fire appearing fo
manifeftly to come from heaven. But it is worth remark-
ing, that perfons killed in this
manner were not always looked
upon as objects of the divine wrath.
That idea would have been too
fevere upon their furviving
friends; and therefore, though
the Mythologifts admitted fuch

* Ad Chariton. p. 258.

a ca-a cataftrophe to be penal, where the party was obnoxious, yet their general doctrine was, that to be ftruck with lightning was a mark of the fpecial favour of JUPITER *·

So far we have endeavoured to point out fome· general principles of MYTHOLOGY. In what follows there is no appearance of any fuch principle, the tranfactions of mankind being indeed too anomalous to be arranged into diftinct claffes. We fhall find however, that when any remarkable fact occurred, the invention of the Mythologifts was always ready to give it a

* Artemidorus Lib. II. cap. 8.

new

new and a magnificent colour-
ing.

We will begin with CECROPS,
the moſt ancient of the Atheni-
an Kings. Of him it is reported that he was half a man and
half a ſerpent; which PLUTARCH*,
and after him EUSTATHIUS †,
ſuppoſe to mean, that in the
former, or at leaſt one half of
his reign, he governed with the
cruelty of a barbarian; in the
other with mildneſs and humanity. It was by an allegory of
the ſame kind according to EU-
STATHIUS ‡, and, if I miſtake

* De ſerâ Num. Vindict. p. 551, Ed.
Franc,
† In Dionyſ. Pericg. v. 390.
‡ In Dionyſ. Perieg, l. c.

not,

not, PLUTARCH alfo, that CAD-
MUS was reported, after his expulfion from Thebes, to be meta-
morphofed into a ferpent ; the
fact being no more, than that
living among the Illyrians, he
loft the humanity of the Greeks
and affumed the favagenefs of
that uncivilized nation.

The metamorphofis of the
people of Ægina from ants into
men is explained by the ancients
themfelves in a manner equally
natural. STRABO * informs us,
that the Inhabitants of that
Ifland, to fave the trouble of
making bricks, were ufed to
dwell in hollows, which they

* L. VIII. p. 576.

dug

dug under ground; a cuftom
that ftill prevails in Wallachia,
and fome parts of Poland, where
dwellings of that fort are called
Limfinks.   The name of ants,
which fo naturally fuggefts itfelf,
and which we can hardly help
applying to a nation living in
this manner, was converted by
the Poets into a ferious fiction,
that the people of Ægina
originated from ants, who afterwards at the interceffion of
ÆACUS were turned into men.
Many other ftories of *Metamor-
phofes* are in like manner founded
upon an analogy, that either
the character, or fituation, of
the perfon had with that of the
bird, or beaft, into which he
was transformed.   LYCAON was
                                    fuppofed

fuppofed, to become a wolf, as refembling that Animal in his favage cruelty. MEROPS,. king of Cos, was faid to be turned into an eagle *, as an emblem, according to ÆLIAN, of his having been a pirate. After TE-REUS had been changed into an *Upupa* by the Mythologifts, the animofity of that Bird againft the Swallow and the Nightingale might poffibly fuggeft the Idea, that PROCNE and PHILO-MELA had affumed thofe fhapes. The anxiety of ALCYONE, and her conftant waiting on the feafhore for the return of CEYX, bore fo great a refemblance to the life of a Kings-fifher, that it

* Euftath. in Iliad. α.

was

was little more than a metaphor
to fay fhe became one. The
followers of DIOMEDE, who died
of famine on fome defert part
of the coaft of Italy, were faid
in like manner to be turned into
Herons, a bird that lives in the
falt marfhes upon fifh. This
probably was the exact fituation
of DIOMEDE and his Argives in
the height of their diftrefs; and
therefore their transformation,
if underftood of this time, was
hardly to be called fabulous.

The power afcribed to the
fountain SALMACIS, of turning
men into women, is thus explained by VITRUVIUS*. That fountain

* Lib. II. cap. 8.

was

was fituated at the very fummit
of the hill, which overlooked the
city of Halicarnaffus; a fpot
originally occupied by the barbarous Carians, who were dif-
poffeffed by a colony of Greeks
from Argos and Troezene. Un-
able to defend themfelves from
this injury, the barbarians endeavoured to revenge it by continual incurfions upon their in-
vaders. At laft however one
of the new fettlers opening a
fhop upon the borders of this
fountain, and having furnifhed
it plentifully with all forts of
wares, the barbarians were allured one after another to fre-
quent it; and in confequence
of that, fays this Author, *e duro*
feroque

feroque more commutati in Græ corum confuetudinem et fuavitatem fuâ voluntate reducebantur. So that the water did not produce effeminacy, according to the common tradition, but only that foftnefs of difpofition which we call humanity.

The imprifonment of MARS by OTUS and EPHIALTES furnifhes another inftance of an hiftorical fact converted into a fable. EUS-TATHIUS informs us*, that OTUS and EPHIALTES were two princes, who by their power and the terror of their arms kept all their neighbours in awe, and obliged them to defift from their mutual

* In Iliad. E. v. 380.

3   invafions

invafions and piracies; which in the language of the Poets was putting MARS in prifon.

It appears to have been a cuftom among the Greeks, in building the walls of any new city, to animate the builders, and give an air of feftivity to the undertaking, by mufic. This at leaft I infer from the manner, in which PAUSANIAS fpeaks of the mufic ufed at the building of the new Meffene by EPAMINONDAS *. Hence we may account for what the Mythologifts report of AMPHION, that the melody of his lyre was fo attractive, that the

* Meffen. p. 345.

F   very

very ftones followed him, and formed themfelves fpontaneoufly into a wall furrounding the city of Thebes; by which, I fup-pofe, nothing more is meant, than that the wall was fo expeditioufly built under his infpection, and to the found of his mufic, as if the ftones themfelves had been animated by it, and arranged themfelves of their own accord in their proper places.

The ftory of DÆDALUS and ICARUS is, I think, well explained by PAUSANIAS *. He tells us, that fails were the invention of DÆDALUS, and firft

* Bœotic. p. 732.

ufed

uſed to facilitate his eſcape from
MINOS, who was only able to
follow him with oars. The
ſucceſs of the invention to the
father and ſon was ſuitable to
the care of the one, and the raſh-
neſs of the other. The father
got ſafe to the deſired port; the
ſon, by carrying too much ſail,
was overſet and drowned.

It is however but a ſmall part
of the Greek Fables that is founded upon domeſtic facts. The
principal of them are ingrafted
upon the ſtories of their four principal Heroes, PERSEUS, BAC-
CHUS, JASON, and HERCULES;
who having been engaged, ac-
F 2   cording

cording to the Poets, in expeditions to diftant or unknown countries, their adventures would of courfe be very confufedly and varioufly related, and muft naturally give great fcope for invention and embellifhment. Had COLUMBUS, DRAKE, and RALEIGH, lived in times, when writing was little practifed, and every thing was preferved in fongs; and when moreover there were no maps to affift the memory; it is eafy to guefs how ftrangely their exploits would have been related; how many miraculous embellifhments they would have received from accident or defign ; and, laftly, how

Geo-

Geographers, in fubfequent times, would have been perplexed to fix the fcene of them. If this would certainly have happened where there was a real foundation of truth, it clearly demonftrates, that there may be fuch a foundation, even where the additional circumftances are impoffible or contradictory. A rational enquirer will therefore reject only that part of the ftory which is varioufly related; and admit that, if not manifeftly abfurd, in which all agree. Though he doubts the miraculous adventures of thofe heroes, he will admit their exiftence, efpecially as the traditions, which atteft

F 3　　　　this,

this, have alfo handed down to us an account of their parentage and their defcendents.

PERSEUS, if admitted to be a real perfon, fhould feem to have been a piratical adventurer, who, having collected together a band of followers, was fortunate enough to fucceed in plundering fome rich city or temple on the coaft of Barbary. His other exploit in Palæftine might perhaps be the refcuing of ANDROMEDA not from a fea-monfter, but from another pirate like himfelf. After this he invaded Argos, pretending to be the Grandfon of ACRISIUS, whom he expelled, and reigned in in his ftead. The ftory of his mother DANAE being thruft out to fea in a fmall boat, with him an infant, and their being afterwards found by fifhermen, and carried by them to the neighbouring ifland of Seriphos, is neither impoffible, nor at all unfuitable to the inhumanity of thofe ages.

BACCHUS, that is, the GRÆCIAN BACCHUS, was an adventurer contemporary with PERSEUS, but of a very different kind. He appears to have been a real perfon, from the difhonourable circumftance of his having been vanquifhed in battle by

F 4   PER-

PERSEUS and the Argives. PAU-
SANIAS, who has recorded this
fact, faw himfelf the fepulchres
of feveral of the female warriors *. He appears to have been
equally unfuccefsful againft Ly-
CURGUS, when he fled for refuge
into the bofom of the fea †; that
is, to his fhips. Facts, fo little
calculated to give luftre, were
not likely to have been feigned
of a Deity. Againft PEN-
THEUS indeed he had better fuccefs, through the indifcretion of
that prince in coming either difguifed, or without a fufficient
efcort, to be a fpectator of his

* Corinthiac. p. 155.
† Iliad. Z. 135.

private

private orgies. He feems to have
been the firft perfon who thought
of attaching the populace to himfelf by falfe miracles, and of ce-
menting his connection with them
by nocturnal affemblies and re-
vels. By thefe contrivances he
appears to have grown in the
end formidable to government in
feveral countries, though never
fufficiently powerful to affume it
himfelf.

The word *belief* is too ftrong
to be given to any notions that
can be formed of a fubject fo obfcure. But the probability of
what I have juft now mentioned
is rather greater than what is a
ad-

adventurer, who was a man of full age when he came to Thebes, fhould be the fon of SEMELE, though it ferved his purpofes no doubt to pretend fo. Neither is it neceffary to fuppofe, that he was the fame perfon who conquered India, whofe character and name he might affume. In what manner he made his way may be gueffed in part from what is recorded in LIVY * of the young man, who firft attempted to introduce the Bacchanalian rites into Rome. The detail of this confederacy feems to me to be the beft key to the arts of the Theban BACCHUS. It is not im-

* Lib. xxxix. 8.

poffible

poffible indeed that he might have penetrated into India, through a country, in which there were neither fortifications, perhaps, nor great cities : But if he did, it is more likely that he fhould have been overpowered and expelled by a frefh confederacy of the inhabitants, than that his retreat fhould have been purely voluntary.

The ftory of Jason and his expedition to Colchis is fuller of miracles and contradictions than almoft any part of the Græcian Mythology ; and therefore we ought not to wonder that it fhould be confidered by many of

the

the moderns as an abfolute fiction, deftitute of any hiftorical bafis. Yet the ancients all admit it as a fact; their Chronologers fix the very year in which it took place; and their Geographers, with equal gravity, fpecify the port from which they fet fail, and thofe which they touched at in their voyage out and home. And as to the perfons concerned in the expedition, nothing can be more particular than the account given by the Grammarians of their parentage and the places of their refidence.

The object of their voyage, as the poets reprefent it, was truly ridi-

6 *Pindari. Pyth. IV.*

ridiculous ; but, as explained by hiftorians, was every way ade-quate to the difficulty of the un-dertaking. The fable of the golden fleece, according to STRA-BO *, took its rife from the method ufed by the inhabitants of Phafis to entangle and collect the gold duft wafhed down from the hills ; which was by placing acrofs the rivers a number of fheepfkins with the fleeces ad-hering to them.* And this is confirmed by APPIAN *, who intimates that POMPEY the Great, after the defeat of MITHRIDATES, made himfelf an eye-witnefs of

* L. XI. p. 499.
† Mithridat. p. 242. Ed. H Steph.

e

the fact. It was natural therefore for the Greeks to confider
the country of Colchis as a fort
of Peru, efpecially when the
riches of it were magnified, as
no doubt they were, by the
marvellous reports of travellers.
It was not therefore a *fingle fleece*
that allured them, but the conqueft or general plunder of the
country. Now this, it is obvious, was not to be effected by
fo fmall a number of warriors
as one and fifty, which is the
higheft number mentioned in
any of the lifts : and we are
obliged therefore either to reject
the ftory entirely, or to fuppofe
with CHARAX, an ancient Grammarian,

marian, that, inftead of one and
fifty mariners, each of thefe
fuppofed mariners was captain
of a feparate fhip, if not com-
mander of a little fleet. It
fhould feem from STRABO * that
they at firft penetrated far into
the country, which the fudden-
nefs of the attack will very well
account for; but their precipi-
tate retreat from Colchis, the
formidable fleet fent after them
by the natives, their being compelled to take a different courfe
in their return, and the little
we hear of the Argonauts afterwards; all this clearly fhews
that their fuccefs was not per-

* L. I. p. 77.

manent;

manent; but that they were difgracefully defeated and difappointed of their booty. Had it been otherwife, I think we fhould have heard more of the *Golden Fleece* after its arrival in Theffaly, than barely what Apollodorus fays [*], that it was prefented to Pelias.

What I have already faid will fufficiently obviate one of the arguments, with which Mr. Bryant has attempted to annihilate the hiftorical bafis of this ftory. He is right in faying, that *the crew of a little Bilander* [†] could

---

[*] L. I. c. 27.
[†] Analyfis, Vol. II. p. 487.

not

not atchieve fo many exploits;
defeat armies, build cities, and
leave feveral colonies behind
them. This is a point given up
by all attentive and critical en-
quirers, both ancient and modern.
Nor is the conqueſt of Peru by
PIZARRO, though effected with
a mere handful of men, at all a
fimilar cafe. Yet it is far from
impoffible, that the Mythologiſts,
to render the ſtory more intereſt-
ing and furprifing, may have
dropped all mention of the *viles
animæ*, that conſtituted the bulk
of the army. And this is the
more probable, as we find the
fame thing practifed in refpect
to HERCULES, who is often re-

G prefented

prefented as having atchieved by perfonal ftrength, what he only did at the head of his troops. Thus of the defeating the Minyæ Euripides fays * :

Ὃς εἰς Μινύαισι πᾶσι διὰ μάχης μο-
λὼν

Θήβαις ἔθηκεν ὄμμ᾽ ἐλεύθερον βλέπειν.

Whereas Diodorus † exprefsly tells us, that he was not the fingle actor in this exploit; but accompanied by all the young men of Thebes.

Mr. Bryant ‡ infifts ftrongly upon the contradictory accounts

* Herc. Fur. ver. 220.
† L. iv. p. 256. Ed. Wefleling.
‡ Analyfis, Vol. II. p. 484. feq.

given

given by different authors of this
expedition : which in his idea
entirely deftroy the credit of the
ftory.   But this furely is inevi-
table in a matter which the
poets, who firft recorded it, collected only from report, and,
where that was imperfect, fupplied the deficiency from fancy
and conjecture.   Before the par-
ticulars of Mr. BANKSES voyage
round the world were commu-
nicated to the public, feveral
different reports were circulated
in refpect to the countries difcovered and vifited ; which reports,
if fuffered to go down to pofterity   without   contradiction,
would have formed the bafis of

G 2   fo

fo many different hiftories. Yet
I think pofterity would have
reafoned ill to have denied the
exiftence of that gentleman, becaufe fome had infifted that he
returned home by the North,
and others by the South, Coaft
of New Holland. And why
might not the Greeks in like
manner have full evidence of the
exiftence of JASON, TIPHYS,
ANCÆUS and others; and that
they failed upon an expedition
to Colchis, and returned; without knowing exactly the rivers
which they failed down, or the
feas and countries which they
traverfed? The one might be a
matter of public notoriety, but
the

the other required accurate in-
formation from the mouth of the
adventurers themſelves.

But Mr. BRYANT * contends,
that the ARGO muſt be a memo-
rial of the *Ark*, becauſe it is ſaid
by ERATOSTHENES † to have
been *the firſt ſhip ever built* ;
which he truly obſerves to be in-
conſiſtent with what the Greek
Poets and Hiſtorians have related of the ſtill earlier voyages
of CADMUS and DANAUS, to mention no more : and from this in-
conſiſtency he again infers, that
they knew not the origin of their
own traditions. But it ſhould be

* Analyſis, vol. II. p. 493. ſeq.
† Aſteriſm. p. 13. ed. Oxon.

G 3   obſerved,

obferved, that the ancient writers are far from being unanimous in reprefenting the ARGO as *the firft fhip ever built*. DIODORUS SICU-LUS tells us *, that it was the firft fhip that had ever been built *of fo confiderable a fize* ; and PLINY the Naturalift †, that it was the firft *long* fhip. If we only fuppofe, that the ARGO was *the firft fhip*, of which any memory or tradition had been preferved, *that failed from Greece* upon a diftant and hazardous expedition, we need not be furprifed, I think, to find, that in time it came to be confidered, in the popular

* L. IV. p. 285. Ed. Wefleling.
† L. VII. c. 57.

MYTHO-

MYTHOLOGY of Greece, as *the firſt ſhip that was ever built.*

Mr. BRYANT further ſays *, that the Argonautic hiſtory muſt have had its origin in ſome country ſouth of Greece, becauſe the *conſtellation* ARGO is *not viſible in ſo northern a latitude.* But this argument I apprehend is much more forcibly applied in another place † to combate the *ſuppoſition* of Sir ISAAC NEWTON, that the ſphere, in which the *conſtellation* ARGO had a place, was conſtructed by CHIRON for the uſe of the Argonauts. To make it of any weight in the preſent queſ-

* Analyſis, vol. II. p. 497.
† Ibid. vol. II. p. 479.

tion,

tion, Mr. Bryant fhould have fhewn, that the *conftellation* Argo was not vifible in any country inhabited by Greeks, or where the language, and hiftory, and fables of Greece, were current. On the contrary, he allows him-felf *, that it was vifible in Rhodes, where Hipparchus is known to have made fome of his obfervations; and in Cnidus, the birth-place and refidence of the famous aftronomer Eudoxus, whofe defcription of the cœleftial phænomena Aratus is faid to have copied. This argument therefore being put out of the way, I fee no reafon for attri-

* Analyfis, vol. II. p. 497.

buting

buting the groundwork of the ſtory to any nation but the Græcians, who claim it. And this is further confirmed by the word *Argo*, which is evidently of Greek origin, being formed from the adjective αργος, *ſwift*, by the ſame analogy as Γοργω, Κελαινω, Καλλιςω, Αριςω, and, I believe, ſome other proper names are from their kindred adjectives.

Before I quit the ſtory of JA-SON, I will juſt obſerve, that there are two fables connected with it, which admit of no very difficult explanation. The HAR-PIES, who were uſed to come ſuddenly, and carry off the food
that

that was set before PHINEUS,
were probably pirates, who land-
ed every now and then to plunder the coast ; till finally driven
away by the two winged sons of
BOREAS, who commanded a part
of the Argonautic fleet. Perhaps the story of TANTALUS,
starving in the sight of his food,
may denote, in like manner, the
sudden incursions of robbers, who
laid waste his country just before
harvest. But to wave this. The
other miracle, to which I alluded,
was, *the speaking keel* of the ARGO,
which I take to have been such
another juggle as that of SER-
TORIUS, pretending to receive di-
rections from heaven through
the

the means of a doe : which the Spaniards readily believing, obeyed his orders with a blind submission *. But *the speaking keel* might have impofed upon an affembly of mariners in a more enlightened age; nor did it require any thing to carry on the impofture, but the faculty of fpeaking inwardly, without opening the lips or teeth, in the manner of the ἐγγαςριμυθοι.

The ftory of HERCULES is more intricate and perplexed than that of any ancient hero whatever; at the fame time that there is very ftrong evidence of fuch a perfon

* Plutarch. Vit. Sertorii.

having

having exifted in Greece, and
performed many not incredible
exploits. The old *hiftorical*
Mythologifts feem perfectly a-
ware of this, and take care to inform us, that the actions attributed to HERCULES are not all
to be underftood of the *Theban*
HERCULES, but fome of the *Æ-*
gyptian, and others of the *Ty-*
rian; and that all thefe have
been blended together by the
fabulous Mythologifts into one
feries of events. As Ηρακλης has
ftrongly the appearance of a
Græcian name, it feems at firft
view difficult to conceive, how
fuch a name fhould have been
borne by any one in either of
thofe

thofe nations, whofe language
was very different, and who were
very unlikely to borrow from
the Greeks in that very remote
period, in which the Tyrian and
the Ægyptian chieftain are placed. Nor is it all eafy to comprehend, how the actions of one
man fhould be imputed to another, who lived in a remote country above a thoufand years later.
Yet both thefe fuppofitions will
perhaps be rendered credible by
premifing a few reflections. We
may recollect then, that among
the ancients it was no fingular
piece of vanity for princes to affume, or fuffer their flatterers to
confer upon them, the name of
fome

fome Hero or Demigod, whofe character might be fuppofed to have ever fo flight a refemblance to their own. To fay nothing of COMMODUS, the Roman HER-CULES, whofe caprices are no ftandard of ancient cuftoms: HYGINUS, if I miftake not, fome-where fays, that CARANUS, the firft of the Argive race who fet-tled in Macedonia, affumed the name of his anceftor HERCULES. The title of BACCHUS was alfo much coveted. It was affumed by PTOLEMY AULETES, king of Ægypt * ; and among the Ro-mans by MARC ANTONY and CA-

---

* Diod. Sic. Lib. I.

LIGULA.

LIGULA *. The northern traditions alſo inform us, that ODIN, from whom their preſent hiſtory begins, had originally another name, but afterwards aſſumed that of a more ancient ODIN, of whom no other circumſtance is now remembered. In like manner the Ηρακλης of the Greeks, who was originally called Αλκαιος, or Αλκειδης, we may ſuppoſe, aſſumed, or was complimented with, the name of Ηρακλης, from the reſemblance of his exploits to thoſe of a more antient HERCULES, well known in that age. Hence in proceſs of time, when the memory of the former was

* Euſtathius in Iliad. I.

grown

grown obfolete, his actions and adventures were attributed to the latter. But whence comes it then that this foreign hero had a Græcian name, as Ἡρακλης undoubtedly is, and that not merely a name of Græcian termination, like Δαρειος, Φαρναβαζος, and many others, but compounded of two genuine Greek words, by the fame analogy as Diocles and Athenocles, two names that we meet with in Polyænus? This is indeed a hard queftion to anfwer, unlefs we fuppofe it to be a tranflation of the Ægyptian word, as Heliopolis is a tranflation of On, and Panopolis of Chemmo. Pausanias * mentions

* Phocic. p. 836.

one

one MACERIS, who was furnamed HERCULES by the Ægyptians and Libyans *.

If therefore the fon of ALCMENA affumed the title of a more ancient hero, it is not to be wondered that the itinerant poets and rhapfodifts of Greece fhould afcribe to their own countryman all the brilliant actions of his namefake : when even grave hiftorians are fome-times mifled by the fimilitude of names, attributing to the younger AFRICANUS, for inftance, what was atchieved by the Elder. Thus then it is poffible that the diftant expeditions to Africa, Spain, Italy, and India, may

* Phocic. p. 836.

H belong

belong to the older HERCULES,
and that the theatre of the Græcian chieftain extended no further than from Greece to Lydia
and the Euxine. The words of
ARRIAN are very appofite. "I
" apprehend," fays he *, " that
" the HERCULES worfhiped by
" the Iberians in Tarteffus (near
" which are the fo-called pillars
" of HERCULES) was the Tyrian
" HERCULES † ; becaufe Tartef-
" fus was built by the Phœni-
" cians, and the temple of HER-
" CULES there is conftructed, and
" the facrifices alfo performed,

* De Expedit. Alexandri. II. p. 88.
Ed. Gron.
  † See alfo Appian. Iber. p. 256. Ed.
H. Steph.

" after

" after the Phœnician manner.
" As for GERYONES, againft
" whom the Argive HERCULES
" was fent by EURYSTHEUS, to
" drive away his kine, and bring
" them to Mycenæ, HECATÆUS
" the hiftorian fays, that he had
" no connexion with the country
" of the Iberians ; and that HER-
" CULES was not fent to any
" ifland called Erythea in the
" Atlantic ocean ; but that there
" was a king, called GERYONES,
" in Epirus, near Ambracia and
" Amphilochia ; and that HER-
" CULES drove away the kine from
" thence, being itfelf no mean
" labour. With refpect to which
" I know myfelf, that Epirus is

H 2   " a good

" a good pasture country, and
" breeds excellent kine ; and I
" think it not improbable, that
" the fame of the kine of Epirus,
" and the name of such a king as
" GERYONES, might be carried
" to EURYSTHEUS, but that he
" would never have known even
" the name of the king of the
" Iberians, situated as they are
" at the extremity of Europe, or
" the quality of their cattle." So
far ARRIAN ; and in support
of this notion we may observe,
that though the son of ALCMENA
is mentioned in general terms, as
having cleared the seas of pirates,
not one of the Græcian writers
has specified so much as a single
exploit

exploit of that kind. SALLUST
alſo, in his hiſtory of the Jugur-
thine war, mentions an African
tradition, that HERCULES died in
Spain ; and that, after his death,
conſiderable bodies of Medes,
Perſians, and Armenians, which
had compoſed part of his army,
came over and ſettled in Africa ;
whereas the army of the Argive
chieftain conſiſted of Greeks, and
principally, according to PIN-
DAR *, of Tirynthians. When
therefore this laſt-mentioned poet
affirms, that HERCULES brought
the wild olive-tree (Kotinos) to
Elis from the country of the

* Ol. X. 40. Iſthm. VI. 40.

H 3   Hy-

Hyperboreans*, which the learned GESNER † suppofes to be Portugal, we ought, I believe, to refer the former part of the affertion to the Græcian HERCULES ; and the latter, that is, the expedition to Portugal, to the Tyrian, who might alfo be the perfon that fubdued CACUS the robber, and fought with ERYX.

But how then fhall we reconcile this with what SYNESIUS tells us ‡, that in the public regifters of the city of Cyrene was recorded, till the invafion of the Barbarians, a lift of defcendents in

* Ol. III. 25.
† De Phœnic. Navig. Præl. II. § 8.
‡ Opp. p. 302.   Ed. Petav.

regular

regular fucceffion from HERCU-
LES to himfelf? Does not this
imply, that the Græcian HER-
CULES vifited Africa, and founded Cyrene? Yet on the other
hand PINDAR, who wrote two
Odes * to ARCESILAUS, king of
Cyrene, gives not the leaft hint
of HERCULES having been there,
and much lefs of his having
founded that city. I prefume,
therefore, that SYNESIUS muft
mean only, that Cyrene, having
been fettled from the ifland of
Thera, which was a Spartan colony, the firft fettlers brought
with them, and kept memorials
of their defcent from the kings of

* Pyth. IV & V.

H 4   Sparta

Sparta and their progenitors as far back as Hercules. And this I think will tally fufficiently with what Synesius fays, at the fame time that it accounts for the filence of Pindar.

If we fuppofe, that the fon of Alcmena never failed to Spain or Africa, it will ftrike off from the lift of his labours the acquifition of the Hefperian apples. Of the remaining ten (for the conqueft of the triple Geryones, that is, Geryones and his two brothers, has been already difcuffed), of thefe, I fay, the killing of the Nemeæan Lion may be literally true; as it is

far

far from impoffible, that a ftrong
active man, if lucky enough to
give a lion a ftunning blow on
the head with a heavy club, might
eafily feize and throttle him, before he recovered from his ver-
tigo. PAUSANIAS even mentions
one POLYDAMAS *, who maftered
and killed a lion without any
weapon at all. The fetching of
CERBERUS from hell, PAUSANIAS †
explains to mean the deftroying
of a great ferpent, that inhabited
a chafm of the earth at Tænarus;
which is ftrongly countenanced
by what EUSTATHIUS tells us
from other authors (for he does
not appear to have read the tra-

* Eliac. II. p. 469.
† Lacon. p. 275.

vels

vels of Pausanias), that Cerberus was reprefented by the
ancients with the head of a ferpent.   Nor was this an atchieve-
ment unworthy of Hercules.
Serpents are found at this day in
the Eaft Indies of an enormous
length, from fifty to one hundred and fifty feet : who, except
when gorged with food, are
greatly an overmatch for any
animal.   Such a one it probably
was that the Roman army deftroyed, though not without the
help of a catapult, near the river
Bagrada*. A ferpent of this fize,
lying at the mouth of a tremendous chafm, might naturally be

* A. Gellius, VI. 3.

called

called *the Dog of Hell*, as the hydra, a fuppofed ferpent, is called by EURIPIDES *the Dog of Lerna* *.

Two or three of the labours of HERCULES appear to be hifto-rical facts, difguifed by allegory. The ftory of the HYDRA related probably to fome crew of Banditti, who fheltered themfelves in the woods and, though defeated and driven from place to place, continually returned, till IOLAUS, fetting fire to the woods, compleated their deftruction. The STYMPHALIAN birds muft alfo be allegorical, if APOLLODORUS† had

* Herc. Fur. v. 420.
† Lib. II. p. 108.

any

any foundation for faying, that they took refuge in the woods, to avoid being the prey of *wolves*. As they are reported to have deftroyed the grain (καρπ8ς *) it is not unlikely, that they might be diftreffed people driven from their own houfes by fome powerful enemy, called *wolves* in the fable, and reduced from neceffity to become robbers in their turn. The ftory of their fhooting of their wing-feathers like arrows indicates, I think, the nature of the weapons which they ufed. The ftory concludes with their being driven from the woods by the terror of a brazen rattle, af-

* Diodorus Sic. l. IV. c. 13.

ter

ter which HERCULES deſtroyed them with his arrows. This may ſignify, that he drove them into ſome ambuſcade by a falſe alarm in the oppoſite quarter: made perhaps by the claſhing of ſhields and ſpears.

The STAG run down by HERCULES was probably a man of the name of *Elaphus*, it being common among the Greeks to give the names of animals to men. Thus we find in their writings perſons called Λεων, Κριος, Αλωπηξ, Μυς, Λυκος, and therefore why not Ελαφος ? As the ſtag is ſaid to have been conſecrated to DIANA, the man I ſuppoſe was one

of

of her priests, who had been guilty of some crime. HERCU-LES was desirous to take him without hurting him ; but not being able, after a year's pursuit, according to APOLLODORUS *, to effect this, at last lamed him with an arrow, and delivered him, still breathing, to EURYS-THEUS.

As the son of ALCMENA was never unattended with something like an army, his expedition against the AMAZONS, and that against DIOMEDE, king of Thrace, carry in them nothing marvellous. The sole difficulty here

* Lib. II. p. 103.

arises

arifes from the object, which hardly feems to deferve a military equipment. Yet it is not diffi-cult to conceive, that if EURYS-THEUS was folicitous, as we find he was, to improve the breed of his cattle by bringing Heifers from Epirus, he might think it a matter of equal importance to mend the breed of his horfes by bringing ftallions, or mares, from Thrace. The *girdle* of ANTI-OPE may feem to have been a very trivial confideration ; but if by *girdle* was meant a tract of coaft, as in the following verfe of APOLLONIUS *,

Ακτῆς Θρηικίης ΖΩΝΗΣ ἔπι τηλε-θόωσαι,

* Argonaut. I. 29.

it

it will not be difficult to difcover the end which Eurystheus had in view; as the coaft adjoining to the Amazons' country, perhaps under their goverment, was precifely the place, if not the only place, where iron was forged and manufactured in thofe remote times; and it was fo much the employment and fupport of the inhabitants, that Apollonius tells us *, they entirely neglected agriculture.

Of the *twelve* labours of Hercules the only material one unexplained is his cleanfing of the Augean Stables, that is, the

* Argonaut. II. 1004.

fences,

fences, in which AUGEAS fecured
his cattle by night, and which
are reprefented as being over-
filled with dung, fo as to require
immenfe labour to remove it.
This however is plainly nonfen-
fical, as AUGEAS no doubt employed cowherds enough (and
very few we know would be fufficient) to remove the dung out
of the ftalls from day to day.
DIODORUS SICULUS tells the ftory
indeed in a peculiar way. According to him *, this was a work
of no great difficulty; but he
fuppofes that HERCULES, by way
of degradation and ignominy,
was enjoined to cleanfe thefe fta-

* Lib. iv. c. 3.

I   bles

bles of the dung, without any affiftance. But this is inconfiftent with what all the Mythologifts fay, that Augeas ftipulated to give Hercules a large reward; which plainly fhews the tafk, whatever it was, to have been of fome importance, and too great for the number of labourers engaged in his fervice. Pausanias relates[*], that the country was fo covered with dung, as to have been rendered incapable of cultivation. And this indeed may have fome foundation of truth in that warm climate, fince, according to Theophrastus, dung without

[*] Eliac. I. p. 377.

water

water burns the ground. But a further folution of this difficulty is, I think, to be found in PLINY, who informs us *, that AUGEAS, king of Elis, was the firft who practifed the manuring of lands with dung in Greece. He probably therefore had difcovered how much his lands fuffered for want of water, and employed HERCULES to remedy the inconvenience : which the latter effected, not by any perfonal labour, but by fetting his foldiers to turn the courfe of a river, or lay dams acrofs it, fo as, one way or other, to overflow the country. The fervice, it is pro-

* N.H. Lib. XVII. c. 6.

bable,

bable, was beyond expectation; as HERCULES demanded, in return for it, the half of his kingdom.

Of the explanations of ANCIENT MYTHOLOGY here given, no inconfiderable number has been handed down by the ancients themfelves, and therefore may be confidered as ftanding upon fome ground of evidence furnifhed by authors and records now loft; the remainder are merely conjectures, upon which criticifm has its full fcope. Confidering the fable as an *Ænigma*, the thing required is, to find fome probable fact, bearing fuch

a rea refemblance to it, as that the fable fhall appear to be only a natural and eafy allegory of the truth. The compleat analogy between them is the only evidence of which the cafe will admit.

DISSERTATION II.

AN

EXAMINATION

OF

SIR ISAAC NEWTON'S

OBJECTIONS

TO THE

CHRONOLOGY OF THE

OLYMPIADS.

I 4

DISSERTATION II.

AN EXAMINATION OF SIR ISAAC NEW‚
TON'S OBJECTIONS TO THE CHRONO-
LOGY OF THE OLYMPIADS.

THE Regifter of the Olym-
pic Games, which is the
great bafis of the Græcian Chronology, was, I believe, admitted as authentic by all the *Lite-
rati* from the revival of letters
to the time of Sir ISAAC NEWTON.
That moft eminent philofopher,
after having inyeftigated with
fuccefs the laws of the material
world, feems to have imagined,
that the fame mathematical knowledge,

ledge, which had enabled him to do this, would be equally ferviceable to him in unravelling the difficulties of ancient hiftory, and reconciling the difcordant epochas of different nations. In the profecution of this defign he has attempted to fhew, that almoft all nations have been fond of carrying back their origin to too remote a period; and with that view have falfified their chronology; in particular, that the Græcians, actuated by the fame foolifh vanity as the reft of mankind, have endeavoured to impofe upon the world a fictitious feries of Olympic victors.

I muft here take the liberty to obferve, that the difpofition

of

of the Græcians to magnify their antiquity, is a thing entirely deftitute of proof, any further than it is proved by Sir ISAAC's own book. It ought not therefore to be confidered as a principle of argument, being indeed the conclufion to which his arguments tend : for which reafon I fhall take no further notice of it at prefent; but proceed to confider the probability of his grand pofition, that the firft part of the feries of Olympic victors is purely fictitious, neither the victors, nor the games themfelves, having at that time had any exiftence.

Now in order that we may the better judge, what degree of credit

dit is due to the lift of Olympiads, I will firft ftate what it
was ; with the external evidence
in its favour. The principal
witnefs is PAUSANIAS, who, in
the time of the emperour MAR-
CUS ANTONINUS, travelled over
Greece in fearch of antiquities,
and was particularly attentive to
thofe at Elis. He appears to have
himfelf examined the regifter of
the *Olympionicæ*; for fpeaking
of two ftatues erected to atone for
fome unfair practices of a Rhodian wreftler, he concludes thus *.
" The Elean guides fay further,
" with refpect to thefe Athletes,
" that it happened in the 178th

* Eliac. I. p. 432.

" Olym-

" Olympiad, that EUDELUS re-
" ceived money from PHILOS-
" TRATUS, which PHILOSTRATUS
" was a Rhodian. This ftory I
" found contradicted by the re-
" gifter of the *Olympionicæ* kept
" by the Eleans ; for in that re-
" gifter it is, that in the 178th
" Olympiad STRATO of Alexan-
" dria in the fame day carried
" away the prize of wreftling,
" and of the Pancratium." In another place *, he tells a remarkable ftory of one LICHAS, a Lacedæmonian, who, at a time when
the Lacedæmonians were exclud-
ed from contending in the games,
entered his chariot in the name

* Eliac. II. p. 454.

of

of the people of Thebes, and having won the race put a garland upon the driver's head with his own hands. This the Eleans refenting ordered him to be fcourged ; which produced an invafion of their country on the part of the Lacedæmonians. " The war being at an end," fays PAUSANIAS, " LICHAS erected his " ftatue here ; but the records " of the Eleans fay, that it was " not LICHAS, but the people of " Thebes, that obtained the " prize." In a third paffage he fays *, " the victory of the fons " of PHIDOLAS is regiftered in the " records of the Eleans as hav-

* Eliac. II. p. 484.

" ing

" ing happened in the 68th " Olympiad, *and not before.*" The text has it περὶ ταύτης, by miftake for πρὸ ταύτης. From thefe three paffages it feems plain, that he had infpected the regifter himfelf. I fay, regifter, in the fingular number, though PAUSANIAS commonly fpeaks in the plural, τὰ Ἠλείων γράμματα. For if he had meant private memorials kept by the Eleans, he would hardly have faid fo confidently, as he does in the laft quoted paffage, that the victory of the fons of PHIDOLAS was regiftered as in the 68th Olympiad, *and not before.* Befides, in fpeaking of the 104th Olympiad, at which the Arcadians prefided, he

fays

says *, " the Eleans do not enter it ἐν καταλόγῳ Ὀλυμπιάδων," plainly intimating, that it was one single catalogue, and the public record of the Eleans. If any doubt should remain, it seems to be removed by what he says of EUANORIDAS the Elean †, that having obtained the prize of wrestling at the Olympic Games, and afterwards becoming one of the *Hellenodicæ*, or presidents, he himself recorded the names of the victors. So that here we see it was not only a public record, but the entries in it were made by the president of the games.

* Eliac. II. p. 529.
† Ibid. p. 471.

I thought

I thought it neceſſary to explain this at large, becauſe we are told by PLUTARCH, in his life of NUMA*, that ſome obje&ed to the liſt of the Olympiads, as being publiſhed by HIPPIAS the Elean rather late, and from no very certain materials. It appears from PAUSANIAS, an eye-witneſs, that he had the beſt materials imaginable; and being a man in high eſteem among the Greeks, and greatly en-truſted by his own countrymen, though ridiculed by PLATO, it is not to be ſuppoſed, that he would publiſh a mere invention

* P. 130. Ed. Bryan.

K   of

of his own as a copy of their records. Befides, PAUSANIAS, who was upon the fpot, and infpected them, gives no hint of their differing materially, or even at all, from the lift commonly received. And we may obferve, that XENOPHON, who lived in the country of Elis, and in the very next generation after HIPPIAS, cites one Olympiad numerically *, and probably therefore from his lift. So that Sir ISAAC is hardly correct in faying †, that the reckoning by Olympiads was not in ufe among the Greeks till after EPHORUS.

* Hellen. I.
† Chronol. p. 47.

Let

Let us next enquire, whether this regifter was kept from the very beginning of the Olympiads. And to this alfo PAUSANIAS bears teftimony; for he fays *, that from the time that there are uninterrupted memorials of the Olympiads, the firft prize given was that of the foot-race, which COROEBUS the Elean won. STRABO alfo mentions COROEBUS as victor in the firft Olympiad †; and ATHENÆUS particularly remarks ‡, that he was by profeffion a cook. The fame PAUSANIAS, fpeaking of a victory gained in a war by the

* Eliac. I. p. 394.
† L. VIII. p. 544.
‡ L. IX. p. 382.

people

people of Megara over the Co-
rinthians, which was prior in
time to the annual Archons of
Athens, fubjoins, " but neither
" were the Olympiads as yet re-
" giftered by the Eleans * ;"
which implies, that they began
to do it foon after; and moft
clearly fhews, that a public regifter was the thing which he
meant; fince no man could fay
with confidence, that no private
memorials were kept before the
time when CoroEBUS was victor.

That fuch a regifter fhould be
at all corrupted, is highly improbable. It is well known how

* Eliac. II. p. 500.

care-

careful the Greeks were of their archives. Thofe of the Athenians were kept in the temple of CYBELE, called METROUM, from whence APELLICON the Grammarian found means to fteal fome of them, which would have coft him his life, as ATHENÆUS informs us *, if he had not fled out of the country. Even the alteration of records was a capital crime among the Rhodians, as we learn from DION †; though in other places, according to Cicero ‡, it was only punifhed by fine. That the Elean records

* L. V. p. 214.
† Orat. XXXI. p. 336.
‡ In Verrem.

K 3   had

had ever been altered, we have no reason to think, as no traces of such a crime are to be found in any ancient author.

But what is the alteration which Sir Isaac suspects, or rather asserts, to have been made? Not merely the erazing of a name here and there, and substituting of another in its stead, but the entire and wilful forgery of about forty Olympiads, that had no real existence *. This bold supposition far exceeds any thing that I recollect in the works of Dr. Bentley, whom a great poet has stigmatized with the epithet

* Chronol. p. 122.

of

of *flaſhing*. But waiving the boldneſs, let us conſider the probability of it. And here two queſtions occur; firſt, what could be the motive? and, ſecondly, how it could be accompliſhed?

The motive muſt be either private or public; private, to gratify the pride of a number of families, by attributing to them Olympic victories which they had never obtained : or public, to increaſe the reputed antiquity of the Olympic inſtitution. Both theſe motives, and the firſt in particular, ſuppoſe that the forged Olympiads were to be imme-

K 4. diately

diately publifhed and circulated, and not to lie dormant upon their regifter; for in that cafe how could the pride of any family be gratified? Suppofing therefore this feries of forty Olympiads, with the names of their refpective victors, to be publifhed, I would afk, as the memory of real victories was always carefully preferved in families, whether the families, to whofe pride this facrifice was made, muft not be perfectly confcious, that no fuch games or victories, as thofe recorded in the forged Olympiads, had ever exifted? And as the victors in forty Olympiads, if all the games were fupplied

with

with fictitious victors, must
have been very numerous, and,
upon the narrowest computation,
could not be fewer than forty,
the number of families entrusted
with this secret must also have
been proportionally numerous,
much too numerous, in my
Judgement, to admit of the forgery passing without detection.

The public motive comes next
to be considered. Now, as the
Eleans, who had the custody of
the Olympic Register, did not
date their origin, as a nation,
from the commencement of the
Olympic Games, but carried it
many generations further back,

I can

I cannot comprehend how their vanity could be materially gratified by proving, that this particular inftitution was fet on foot in the fourth century of their political exiftence, rather than in the fifth or fixth. Or will it be faid, that, by carrying back the origin of the Olympic Games, they in effect carried back the original æra of their own hiftory? Even this could be no object to the Eleans, becaufe their hiftory is fo intimately connected with that of the petty nations adjoining, that whatever heightens the antiquity of any one muft equally heighten that of the others. The whole mafs of

Græcian

Græcian history muft move together. To increafe the number
of the Olympiads, could not
therefore contribute in the leaft
to fet them above the heads of
their neighbours, which is the
common, if not the only, motive, for pretending to fuperior
antiquity. Of the more diftant
and barbarous nations, they either
knew too little to enter into any
competition with them, or elfe
they fhewed their indifference for
this fancied honour, by readily
fubfcribing, as in the cafe of the
Ægyptians, to the claim of pri-
ority, which thefe laft, with great
confidence, and perhaps with
juftice, urged.

We

We are next to enquire into the poſſibility of impoſing ſuch a fiction upon the world. It is evident, that no ſuch impoſition could take place, after the time when HIPPIAS the Elean publiſh-ed the liſt of the *Olympionicæ*, as mentioned by PLUTARCH [*]. Sir ISAAC ſays, that he lived in the 105th Olympiad [†]; and poſſibly he might live till that time; but it appears, that in the time of SOCRATES, whoſe death happen-ed in the beginning of the 95th, which is no leſs than forty years before, he had already attained to great fame, honours, and wealth: and therefore the pub-

[*] Vit. Numæ. l. c.
[†] Chronol. p. 47.

lication

lication of his lift may as proper-
ly be fixed to this time, as to any
other. Now, according to Sir
Isaac's reckoning, the Olympiad, which we now call the 95th,
was in reality the 55th, and the
Olympic Games of confequence
had only been celebrated about
220 years. It is hard, it is impoffible to conceive, that among
the Greeks, who had fo long
had the ufe of letters, who had
many ancient writings preferved
among them, and who were fo
remarkably fond of genealogies,
that they, I fay, fhould have fo
entirely loft all memory of the
inftitution of the Olympiads, as
not to know whether they
had

had lafted 380 years, or only
220. Set the date of the publication by HIPPIAS higher, and
you ftill increafe the difficulty.

This however is far from being
all. In the catalogue of the
Olympiads, every Olympiad had
its particular victors, whofe
countries, as well as their names,
were fpecified. The memory of
an Olympic Victor was fo carefully preferved in Greece, it was
fo precious to the nation, as well
as family of the Victor, that,
when a new lift came out of
forty or more Victors that had
never been heard of before, what
muft have been the aftonifhment

of

of thofe cities and countries upon which this honour was fo gene-roufly conferred by the Eleans? Muft they not have feen through the impofture at once? and, if not at once, how could it efcape detection, perfect and complete detection, for fo many centuries together, in a nation abounding with criticks? For that the lift was publifhed without any chafms we have undoubted proof, becaufe the names of all the Victors in the *Stadium* are come down even to us, and many of them alfo are mentioned by more authors than one, who all agree as to the date of the vic-tory.

The

The impofture, if it took place at all, muft relate entirely to the firft Olympiads; becaufe the nearer it approached to the time of publication, the more certainly it would be detected. Now, the authenticity of thofe Olympiads is ftill further corroborated by a variety of little circumftances preferved in PAUSANIAS, no way neceffary to the fuppofed fraud, and therefore not likely to have been invented. We are told for inftance *, that the firft prize was that of the *foot-race*; that in the fourteenth Olympiad that of the *Diaulos*, or *double foot-race*, was added; that in the eighteenth they re-

* Eliac. I. p. 394.

vived

vived the *Pentathlon* and the game of *Wreftling* ; that in the twenty-third the prize of *Boxing* was inftituted, and in the twenty-eighth that for full-aged horfes. Are not all thefe marks of reality ? and is there any inftance of an impofture being clogged with fo many uneffential circumftances ?

Upon the whole, it is no extravagance to fay, that the lift of the Olympic Victors has an authority equal, if not fuperior, to that of any documents whatever, of the fame kind. It confifted of entries made by a public officer, relative to tranfactions of

L the

the greateſt notoriety, which the
parties themſelves, their rela-
tions, and their country, were
highly intereſted to keep in memory. Greater ſecurity than this
for the integrity of any record it
is impoſſible to have ; and therefore to doubt the truth of it
would be to introduce univerſal
ſcepticiſm.

Having now ſhewn upon what
authority the Olympic Regiſter
ſtands, I come next to conſider
the objections raiſed againſt it
by Sir ISAAC NEWTON ; which
will be found, I apprehend, to
be by no means ſufficient to
overturn the evidence in favour
of

of its general veracity. His firft and principal objection is taken from two lifts of Spartan kings ; that kingdom by a very particular conftitution having been governed by two contemporary kings, the lineal reprefentatives of two brothers, EURYSTHENES and PROCLES, who conquered it. The beginning of the Meffenian war, as calculated by the Greeks, is fixed by PAUSANIAS* to the fecond year of the ninth Olympiad, at which time ALCAMENES of the houfe of EURYSTHENES was king of Sparta, and THEOPOMPUS of the houfe of PRO-

* Meffen. p. 292.

L 2   CLES.

CLES *. From this year, according to the Olympic computation, it was 263 years to the expedition of XERXES, which happened in the firſt year of the 75th Olympiad. In this long interval of 263 years, we find only eight kings of the houſe of EURYSTHENES, excluſive of ALCAMENES, and only ſeven, ſtrictly ſpeaking, of the houſe of PROCLES, excluſive of THEOPOMPUS. The liſt of the *Eurysthenidæ* is as follows: 1. POLYDORUS. 2. EURYCRATES. 3. ANAXANDER. 4. EURYCRATES the Second. 5. LEON. 6. ANAXANDRIDES. 7. CLEOMENES. 8. LEONIDAS. This is

* Meſſen. p. 288.

the

the lift of Kings as given by
PAUSANIAS *. That given by
HERODOTUS † leaves out CLEO-
MENES, becaufe it is not a lift
of Kings, but only of the Progenitors of LEONIDAS, the brother of CLEOMENES. The kings
of the *Proclidæ* family, according to PAUSANIAS ‡, were,
1. ZEUXIDAMUS, who fucceeded
his grandfather THEOPOMPUS.
2. ANAXIDAMUS. 3. ARCHIDA-
MUS. 4. AGASICLES. 5. ARISTO.
6. DEMARATUS, who was de-
pofed, and fucceeded by, 7. LEO-
TYCHIDES. The lift given by

* Lacon. p. 209—214.
† L. VII. c. 204.
‡ Lacon. p. 220, 1.

HE-

HERODOTUS*, differs confiderably from this. According to him their names were, 1. ANAXAN-DRIDAS. 2. ARCHIDAMUS. 3. A-NAXILAUS. 4. LEOTYCHIDES the Firft. 5. HIPPOCRATIDES, whofe fon HEGESILAUS, and grandfon MENARES, not fucceeding to the crown, the next king in fuccef-fion is, 6. LEOTYCHIDES, the fon of MENARES. It is not very material to fettle the difference between HERODOTUS, the older writer, and PAUSANIAS the more diligent antiquary, becaufe though Herodotus leffens the number of kings, he adds one to the number of generations. Now what

* Lacon. VIII. c. 131.

is

is the obfervation of Sir ISAAC upon thefe two lifts? He tells us*, that " by the ordinary courfe " of nature kings reign one with " another about 18 or 20 years " a-piece;" and having laid down this rule, he applies it, among others, to the Spartan kings abovementioned; according to which the interval between the firft and the laft of thofe kings amounts to no more than 140 years; whereas, if we reckon by the Olympiads, it makes, as I faid, 263.

To all fuch reafoning I have one general anfwer; that the

* Chronol. p. 54.

L 4   reigns

reigns of kings not depending upon the common chance of mortality, or upon any fimple and conftant natural caufes, but upon a variety of natural and political caufes, operating in con-junction ; fuch as their own folly or wifdom, the caprice of the multitude, the treachery of their own fubjects, and the in-vafion of foreign powers : all thefe caufes, I fay, render the length of reigns fo uncertain and variable, that though we may form an average of them as we may of any thing elfe, we can-not reafon firmly and folidly upon that average. We cannot ra-tionally fay ; fo many kings of
Perfia,

Perſia, Macedonia, France, or England, reigned, one with another, about 20 years each, and therefore ſo many emperors of Japan did not reign longer. If we take the three laſt kings of France, their reigns amount in the whole to 164 years, which is at the rate of 55 years for each. But the reigns of GALBA, OTHO, and VITELLIUS, three ſucceeding emperors of Rome, did not amount in all to a year and three quarters. What dependence can there be upon a calculation of things that differ ſo enormouſly ? For either we know the hiſtory of the reigns which are the ſub-jeƈt of calculation, or we do not.

If

If we do not, how can we be sure that they did not succeed one another as rapidly as the emperors of Rome just mentioned? or, on the contrary, that the crown did not devolve successively to minors, who enjoyed it peaceably to a good old age, which was nearly the case of the three French kings? If we *do* know the history, then this method of calculation is superseded by positive and substantial evidence. So that in no case is it useful as a medium of proof; and should therefore be rejected, as totally unworthy of attention.

Generations of men, though sufficiently vague, are however a

better

better ground for calculation than the reigns of kings; the one having no other limits than the period of life; whereas the generative faculty does not exift in full force above a third part of that period. Now, in the prefent cafe, the learned and diligent Pausanias, who has probably given us the true lift of both the races of thefe Spartan kings *, (for one of thofe in Herodotus is palpably erroneous), this fame Pausanias has alfo given us a very circumftantial account of their genealogy. Is it not therefore a little extraordinary that Sir Isaac Newton,

* Lacon. p. 209 & 220.

having

having it in his option to calculate this interval of time by the
more accurate method, fhould
chufe to do it by the more vague
and inaccurate? We fhall pre-
fently fee, that had he calculated
by generations, his objection
againft the authenticity of the
Olympiads would not have been
near fo ftrong: and one cannot
therefore help fufpecting, that,
great and candid as he unquef-
tionably was, he was, in this inftance, drawn out of the right
path by a bias, imperceptible to
himfelf, in favour of his own
opinions.

Sir Isaac Newton has ftated
very fairly his method of computation

tation by reigns, and the difference between them and generations. He fays *, that "gene-
" rations from father to fon may
" be reckoned, one with another,
" at about 33 or 34 years a-piece,
" or about three generations to a
" hundred years; but if the
" reckoning proceed by the eld-
" eft fons, they are fhorter, fo
" that three of them may be
" reckoned at about 75 or 80
" years. And the reigns of kings
" are ftill fhorter : becaufe kings
" are fucceeded not only by their
" eldeft fons, but fometimes by
" their brothers; and fometimes
" they are flain or depofed, and

* Chonol. p. 53. 54.

" fuc-

" fucceeded by others of an
" equal or greater age, efpecially
" in elective or turbulent king-
" doms." All this is undoubtedly true ; and being fo, one
does not fee with what proprie-
ty an average, drawn from this
method of computation by reigns,
can be applied to cafes, where
we know, from good authority,
that there was no revolution,
or change of family, but that
the crown defcended peaceably
from father to fon. Suppofe a
calculator was to eftablifh this
rule, that the duration of a fhip,
including accidents from fire
and fhipwreck, was, upon a
medium, 15 years; would this
be

be conclufive with refpect to the duration of any number of fhips, which we might know from good authority to have met with no fuch accident? In like manner, where the crown defcends regularly from father to fon, we have nothing to do with a rule, which pre-fuppofes an interrupted fucceffion. Our bufinefs here is to calculate by generations, where the error, we may confidently fay, cannot be great; whereas in calculating unknown time by reigns, there is no poffibility of gueffing what it may be. If we had no hiftory of the Roman emperors from CÆSAR to CONSTANTINE, and

were

were reduced to the neceffity of calculating by Sir Isaac's average, how extravagantly would it miflead us? For the number of reigns being 44, reckoning the emperors who reigned together only as one, this multiplied by 18, which is Sir Isaac's loweft average, would give 792 for the number of years, whereas in fact they were only 373.

Let us now fee, how the calculation by generations will fuit with the above-mentioned period of 263 years. According to Pausanias *, the generations of the *Eurysthenidæ*, from the fe-

* Lacon. p. 210. 214.

cond

cond year of the ninth Olympiad, were feven in number, exclufive of ALCAMENES, who is the head of the lift; 1. POLYDORUS. 2. EURYCRATES. 3. ANAXANDER. 4. EURYCRATES the Second. 5. LEON. 6. ANAXANDRIDES. 7. LEONIDAS. Thofe of the *Proclidæ*, excluding in like manner THEOPOMPUS, were, according to the fame author *, 1. ARCHIDAMUS, who dying before his father, never reigned. 2. ZEUXIDAMUS. 3. ANAXIDAMUS. 4. ARCHIDAMUS. 5. AGASICLES. 6. ARISTO. 7. DEMARATUS, who, though depofed, was ftill living, and accom-

* Lacon. p. 220, 1.

M   panied

panied Xerxes on his expedition.
Divide 263 years by 7, and the
quotient is 37, with a remainder of 4, which makes 37 years
and a half for each generation.
This, though rather more than
is commonly allowed, is greatly
within the limits of poffibility ;
and therefore the excefs, even if
there were no way of accounting
for it, would not furnifh any
conclufive argument againft the
authenticity of a public record.

But I muft obferve, that the
caufe of this irregularity, is to
be found in the hiftory itfelf.
Anaxandrides, the fixth in our
lift of the Eurysthenidæ, was

fo

fo long without children by his
firft wife, that, according to Pausanias * and Herodotus †, he
was compelled by the *Ephori* to
take another, for fear the race of
Eurysthenes fhould be extinct.
This fecond wife brought him
Cleomenes, his immediate fuc-
ceffor; after which he had three
fons, Dorieus, and Leonidas,
and Cleombrotus, by his firft.
Leonidas therefore was born
when his father was advanced
towards the latter part of the
generative period. It was alfo
late in life before Leonidas fucceeded to the throne; for his

* Lacon. p. 211.
† L. v. c. 39, 40.

half-

half-brother CLEOMENES, who was advanced to it before him, did not become King till his brother DORIEUS was of an age to command a fleet and army, and to fettle a colony. CLEOMENES reigned upwards of twenty years; fo that LEONIDAS, at the time of his acceffion, was probably near forty; and this being twelve years before the expedition of XERXES, LEONIDAS at the time of that event may be confidered as a man of about two and fifty.

A fimilar accident happened about the fame time in the family of the *Proclidæ*. ARISTO, the

the fixth in our lift, had, according to HERODOTUS *, mar-
ried two wives; but, ftill con-
tinuing childlefs, was defirous
of marrying a. third; and caft
his eyes upon the wife of his
friend AGETUS, who, having been
betrayed into. an oath not to
withhold any thing in his poffeffion from the king, furrender-
ed her to him with reluctance.
By this third wife he had DE-
MARATUS, who at the time of
the invafion of Attica by the
Spartans, in fupport of the fac-
tion of ISAGORAS †, was old
enough to be joined in command

* L. VI. c. 62, 63.
† Herod. V. c. 74, 5.

with

with CLEOMENES, and therefore
was probably not lefs than fifty
years of age, when XERXES, near
thirty years after, invaded Greece.

We have here two kings of
Sparta, neither of them born in
the firft youth of his father, and
both of them declining in life at
the celebrated epocha of the battle of Thermopylæ. Taking
thefe circumftances together, we
may fairly, I think, add an
eighth, or nearly an eighth, generation to the feven, for which
we have an inconteftable warrant
from hiftory. And this being
done, if we divide 263, which
is the Olympic number of years,
by

by 8, the quotient will be a little lefs than 33, and it will be half a year fhort of Sir ISAAC's own allowance for a generation, which is between 33 and 34 years. Neither is this allowance greatly too much for the duration of the reigns of Kings, where the fucceffion follows the eldeft fon; for if we examine the genealogies from WILLIAM the Conqueror to his prefent majefty, we fhall find them to be three and twenty in number, excluding, as we ought, either the firft of thefe Kings or the laft; and then, if we divide 713, which is the number of years between 1066, the year of the Conqueft, and

M 4   th

the year 1779, by 23, we shall
find the quotient to be exactly
31. And though the line of
genealogy goes in two instances
through a younger son, that is,
through HENRY the First, and
JOHN of Gaunt, duke of Lan-
caster, which of course adds
something to the length of those
generations: yet in two others
it passes through a sister elder
than the male heir, that is, through
MARGARET daughter of HEN-
RY VII. and the queen of Bohe-
mia, daughter of James I. so
that what is gained by one aber-
ration is lost by the other.

If it be thought probable, that
kings in particular should marry
and

and have heirs before the age of
33 : I answer in the first place,
that both HESIOD and SOLON,
almost the only writers from
whom we can learn the customs
of that age, both these, I say,
have fixed upon thirty, as the
seasonable age for marrying.
The words of HESIOD have been
often quoted * :

Μήτε τριηκόντων ἐτέων μάλα πολλ'
ἀπολείπων,

Μήτ' ἐπιθεὶς μάλα πολλά.

Those of SOLON are still more
apposite. Having divided hu-
man life into stages of seven
years each; he speaks thus of
the fifth † :

* Ἐργ. v. 696.
† Ap. Clem. Alexand. Strom. L. vi.
p. 686.

Πέμπ]η

Πέμπ]η δ᾿ ὥριον ἄνδρα γάμε μεμ-
νημένον ἔναι.

But secondly; though kings
should be suppofed to marry earlier, fome allowance fhould be
made for their wives not being
pregnant immediately; or for
the firft children being daughters;
or, if males, for their dying as
a great proportion of infants does
before they are paft their child-
hood.

So much for the argument
drawn from the ufual time that
kings, upon an average, may
be fuppofed to reign. I come
now to confider a few fcattered
                              paf-

paſſages of the ancient Greek writers, which have been thought to contradict and overthrow the Clympic Chronology.

And firſt, we are told by PLATO *, that the laws of Ly-curgus had been then eſtabliſhed a little more than three hundred years; ἔτη τριακόσια ἢ ὀλίγῳ πλείω. Here if the reading be right, and the author well informed, the argument is concluſive. But as the change of a letter or two frequently makes a great change in the ſenſe, the firſt queſtion to be conſidered is the correctneſs of the text, and whether it can

* In Minoe, p. 567. Ed. F.

6   be

be fo far depended upon, as to overfet the concurrent teftimony of all antiquity. It would be no great alteration to read ἔτη τετρακόσια inftead of τριακόσια, and fuch a miftake might be accounted for feveral ways; either from the firft fyllable of τετρακόσια being obliterated, or elfe thofe two letters τ, ε, might be dropt, from their fimilarity to the laft fyllable of ἔτη; after which the fubftitution of τριακόσια for τρακόσια might very eafily happen. Admitting τετρακοσια to be the true reading, it will fufficiently quadrate with the Olympic Chronology. SOCRATES is fuppofed to have been born in

the

the third year of Olym. LXXVII. that is 307 years after the firſt Olympiad. Add to this 50 years for the age of SOCRATES, at the time when this converſation happened. This makes the number 357. Concerning the time when LYCURGUS flouriſhed there were ſeveral opinions; ſome making him cœval with the firſt Olympiad; others, according to PLUTARCH *, ſetting him many years before it; and others, I preſume, at other intervening periods. Now as we are quite in the dark which of theſe opinions PLATO followed, we are at full liberty to chuſe that, to

* Vit. Lycurg. init.

which

which the expreſſion of ἔτη τετρακόσια ἢ ὀλίγῳ πλέω will moſt
perfectly agree.

Secondly, THUCYDIDES, ac-
cording to Sir ISAAC NEWTON *,
affirms, that " from the time
" the Lacedæmonians had uſed
" one and the ſame adminiſtra-
" tion of their commonwealth
" to the end of the Peloponne-
" ſian war, there were *three hun-*
" *dred years* and a few more."
This difficulty, ſuppoſing the
quotation exact, would be of leſs
conſequence than the preceding ; for it might be got over
without any alteration of the

* Chronol. p. 57.

text.

text. By the words *one and the same adminiftration of the commonwealth* the Hiftorian might mean, not the laws of LYCURGUS, but the new form, which the government affumed in the reign of THEOPOMPUS, by the inftitution of the *Ephori* ; an event which we may fuppofe to have happened in the laft year of Olymp. XI. exactly in the middle of his reign. From this year to the taking of Athens by LYSANDER, there is an interval of 82 Olympiads and one year, that is, of 329 years ; to which fpace the expreffion of THUCYDIDES might naturally and without violence be applied. But in fact THUCYDIDES does not fay what Sir ISAAC im-

imputes to him. The word in all the known copies of the original text * is τετρακόσια, *four hundred*, for which *three hundred* has been fubftituted in the old Latin tranflation, where only it is to be found. This larger number, it is obvious, muft be applied to the laws of LYCURGUS, and there can be no difficulty in applying it, as THUCYDIDES has no where told us, what interval he placed between LYCURGUS and the firft Oympiad.

I fhall not at prefent examine the objection which Sir ISAAC makes † to the Græcian Chro-

* L.1. c. 18. Ed. Duker.
† Chronol. p. 55.

nology,

nology, from the *Lacedæmonian* (not *Meſſenian* *) army having been commanded in the ſecond year of Olymp. X. by one EURY-LEON, the ſixth in deſcent from THERAS, who lived during the invaſion of the *Heraclidæ*, be-cauſe this objection principally affects the time preceding the Olympiads; whereas the preſent matter of diſcuſſion is the ge-nuineneſs of the Olympiads them-ſelves. I proceed therefore to an objection of much more im-portance, which affects the Olympiads only; and, if ſub-ſtantially ſupported, would go a great way towards overthrowing

* Pauſan. Meſſen. p. 296.

N   their

their credit. HERODOTUS, in his lift of the noble youths, who follicited the daughter of CLISTHENES in marriage, mentions one LEOCEDES, the fon of PHIDON. His words are *, ἀπὸ δὲ Πελοπον-νήσ8 Φείδωνος τ8 'Αργέων τυράνν8 παῖς Λεωκήδης, Φείδωνος δὲ τ8 τὰ μέτρα ποιήσαντος Πελοποννησίοισι, κ̀ ὑβρίσαντος μέγιςα δὴ Ἑλλήνων ἀπάν-των· ὅς ἐξαναςήσας τὸς Ἠλέων ἀγω-νοθέτας αὐτὸς τὸν ἐν Ὀλυμπίῃ ἀγῶνα ἔθηκε. The words, as they ftand, can mean nothing but that LEOCEDES was the fon of PHIDON, the tyrant of Argos, and the fame PHIDON, who eftablifhed the Peloponnefian weights and mea-

* Lib. VI. c. 127.

4.   fures ;

fures ; who was guilty of greater excefses than any other of the Greeks, and having eje&ted the *Agonothetæ* of the Eleans pre-fided himfelf at the Olympic Games. The time of the PHI-DON, who ufurped this office, is fixed by PAUSANIAS to the eighth Olympiad *; and with him STRA-BO † in effe&t agrees, by making PHIDON the tenth in defcent from TEMENUS. The Arundel Mar-ble indeed fpeaks of one PHIDON an ARGIVE, who coined money 415 years before the expedition of XERXES ‡; that is 119 years before the firft Olympiad. But

* Eliac. II. p. 509.
† Lib. VIII. p. 549.
‡ Marm. Oxon. I. l. 45.

as

as nothing is faid of his tyranny or invafion of his neighbours, it is poffible that in the Marble, as well as in Herodotus, he may have been confounded with PHI-DON of Corinth, a very ancient Lawgiver mentioned by ARIS-TOTLE *. Or even if the fame PHIDON be meant, the miftake is not greater than fome others which have been found in that famous Chronology. Be this as it may, my argument is not at all benefited by following the computation of PAUSANIAS and STRABO, rather than that of the Marble; becaufe in either cafe, if it fhould clearly appear, that

* Politic. II. p. 35. Ed. Sylburg.

LEO-

LEOCEDES, the fon of this PHI-
DON, courted the daughter of
CLISTHENES, it will equally
prove the incorrectnefs of the
Olympic lift, or rather indeed
the non-exiftence of a great part
of the Olympiads. Sir ISAAC
very juftly places CLISTHENES in
the 47th Olympiad, and his
daughter being courted by the
fon of PHIDON, it follows that he
and PHIDON were nearly equal
in age. Now PHIDON is faid to
have lived in the eighth Olympiad, and CLISTHENES 154 years
later, in the 47th. The confequence evidently is, if they were
really contemporaries, that the
Olympiads muft have been ex-

N 3   tended

tended near 140 years beyond
the truth.

The attentive reader will obferve, that the whole of this argument refts upon one fingle
paffage in HERODOTUS, and con-
fequently falls to the ground, if
that paffage fhould happen to
have been incorrectly tranfcribed;
if it has been interpolated or mu-
tilated, fo as to alter the fenfe.
Now the perfect correctnefs of
any one paffage, confidering the
many corruptions to which books
are liable, is a thing too uncertain to be relied upon, in oppofition to the weight of evidence
in favour of the Olympic Chronology.

nology. In this very paſſage, two of the manuſcripts omit the important word παῖς, which leaves room for ſuppoſing that LEOCEDES might be only the deſcendent of PHIDON. But for my own part I am inclined to believe, that the word παῖς is genuine, and no interpolation; though I think that the paſſage is evidently corrupted. The adverſative particle δὲ comes in very aukwardly and improperly in the ſecond part of the ſentence, ſuppoſing the writer to ſpeak of the ſame PHIDON in both places; and if we ſtrike it out, the repetition of the word Φείδωνος has a poetical air, very unſuitable to hiſtory.

I

I would read therefore Φείδωνος
τȣ̃ Ἀργείων τυράννȣ παῖς Λεωκήδης,
Φείδωνος δὲ ΟΥ τȣ̃ τὰ μέτρα ποιήσαντος, κ. τ. λ. In Englifh: "LEO-
" CEDES the fon of PHIDON, king
" of Argos ; but *not* of that PHI-
" DON who eftablifhed the Pelo-
" ponnefian meafures," and fo
forth. This correction fuppofes
that PHIDON, the father of LEO-
CEDES, though feveral generations later than the enterprizing
PHIDON, ftill poffeffed the fo-
vereignty of Argos : the probability of which it may be neceffary to eftablifh, it not being
generally known, that Argos
continued to be a monarchical
ftate fo long. What Sir ISAAC
NEWTON

NEWTON afferts *, that " between CISUS (the fon of TEME-
NUS) and PHIDON they reigned
not," is a miftake, arifing from
his underftanding the words of
PAUSANIAS in too ftrong a fenfe.
That learned antiquary fays †,
that the Argives, being from the
moft ancient times lovers of equa-
lity and independence, reduced
the power of the kings fo low—
ὡς μηδένι τῶν Κείσυ ἢ τοῖς ἀπογόνοις
ἢ τὸ ὄνομα λειφθῆναι τῆς βασιλείας
μόνον—" *that nothing but the
name of royalty was left to the defcendents of* CISUS." The paffage in the Greek is intricate and

* Chronol. p. 123.
† Corinth. p. 152.

corrupt; nor will it be much mended by reading ὡς ΜΗΔΕΝ μηδένι τῶν κέίσ8 — which would fignify, " *fo that nothing was left to any of the defcendents of* Cisus *and to his defcendents, but the name only of royalty.*" Pausanias, in the following fentence, mentions one Medon, from whom the laſt king of Argos was defcended; and Satyrus, an ancient hiſtorian, quoted by Theophilus, biſhop of Antioch *, fpeaks of Maron as the fon and immediate fucceſſor of Cisus. I would read therefore — ὡς μηδεν ΜΗΔΩΝΙ τῷ Κέίσ8, ἢ τοῖς ἀπογόνοις, ἢ τὸ ὄνομα λειφθῆναι τῆς βασιλείας

* Ad Autolycum. l. II. p. 96. Ed. Wolf.

μόγον

μόνον——"*fo that nothing was left to* MEDON, *the fon of* CISUS, *and his defcendents, but the name only of royalty.*" It is plain from this paffage alone, that the lineal defcendents of CISUS, the fon of TEMENUS, continued to be kings of Argos for fome time; and ARISTOTLE, in his Politicks *, exprefsly fays, that PHIDON the Argive *of a King became a Tyrant*; fo that the kingdom was his by inheritance. But neither was PHIDON the laft to whom it went in fucceffion. For PAUSANIAS, in the fentence immediately following, fays, " that the people [of Argos] being dif-

* L. V. p. 152. Ed. Sylburg.

affected

affected to MELTAS, the fon of
LACIDES, and defcendent of ME-
DON, deprived him of the government entirely. The kingdom of Argos therefore was poffeffed, as an hereditary, though
limited, monarchy, by the fon
of LACIDES; a word, which,
when corrected to LACEDES, as
LEOCIDES in HERODOTUS has
been to LEOCEDES, turns out to
be the fame name, differing no
otherwife than as Μενελεως does
from Μενελας, or LEODAMAS from
LAODAMAS. DEMOCEDES, a name
repeatedly mentioned in HERO-
DOTUS, is a word of the fame
import. Upon the whole then
it feems highly probable, that
the

the LEOCIDES, or LEOCEDES, of
HERODOTUS, was fon to the king
of Argos, whatever the name of
his father was. I fuppofe it to
have been PHIDON, who being
an obfcure perfon, compared with
his anceftor the invader of Elis,
it became neceffary for HERODO-
TUS to caution his readers againft
confounding one with the other.
If the conjecture here propofed,
to which I forefee no material
objection, be admitted, it entirely
removes the Chronological difficulty infifted upon by Sir ISAAC
NEWTON.

A fourth objection to the
Olympic Chronology is taken

7   from

from a paffage in Pausanias *,
where he is fuppofed to fay, "that
Cypselus, king of Corinth, was
the fixth in defcent from Melas,
the contemporary of Aletes,
who got poffeffion of Corinth
when the *Heraclidæ* returned
into Peloponnefus †. The reign
of Cypselus began in the 30th
or 31ft Olympiad; and by this
reckoning Melas muft have lived
only two generations before the
firft Olympiad; whereas his contemporary Aletes was alfo contemporary with Temenus ‡, feven
or eight generations older than
that Period. Admitting there-

* Eliac I. p. 424.
† Chronol. p. 62.
‡ Strabo, L. VIII, p. 597.

fore

fore Cypselus to have been only
the fixth from Melas, we muft
annihilate about five generations,
which are nearly equal to one
and forty Olympiads, of the interval between them. But the
fame Pausanias, from whom
this is quoted, tells us in another place*, that Aletes and
his defcendents reigned at Corinth
for five generations, the laft
being Bacchis, the fon of Prum-
nis : that after him the fo-called
Bacchiadæ reigned there for five
other generations, ending with
Telestes, the fon of Aristo-
demus, who was killed by Ari-
eus and Perantas: after which

* Corinthiac. p. 120.

Corinth

Corinth was not governed by
kings, but by annual magiftrates
of the race of the *Bacchiadæ*, till
the time of their expulfion by
Cypselus." This account of
Pausanias is clear and circumftantial; and from it we learn,
that there were at leaft ten generations between the time of Melas and Cypselus; and how
many more we are not informed.
The word ἑκτον therefore, in the
paffage of Pausanias, quoted
by Sir Isaac, is indifputably a
corruption. It might be altered
to ἐνδεκατον; but a careful examination of the context has
convinced me, that the original word was not a word
of

of number. The paſſage ſtands thus in the editions * : Κυψέλῳ ἐ τοῖς προγόνοις ἕκτον ἦν γένος ἐξαρχῆς Γονέσης τῆς Σικυῶνος, ἐ πρόγονος σφίσιν ἦν Μέλας ὁ Αντάσσε. If this paſſage admits of any ſenſe at all, it muſt be ſomething like the following: CYPSELUS *and his an-ceſtors were in the ſixth genera-tion from* GONUSSA *of Sicyon, and their Progenitor was* MELAS *the ſon of* ANTASSUS. But this, as the reader muſt ſee, is full of ab-ſurdities. GONUSSA is the name of a *place* in the country of Si-cyon ; and not, as Sir ISAAC ima-gined, of a *perſon*. It is more-over a palpable blunder to ſay,

* Pauſan. Eliac. I. p. 424.

O   that

that Cypselus, *and his ances-tors,* were in the fixth genera-tion from any body; for if he was in the fixth, his father muft have been in the fifth; and his grandfather in the fourth. It appears probable to me, that EKTON was formerly EK ΓON, and that Γονᾶσης does not occu-py its proper place, but fhould follow the præpofition ἐκ, the whole ftanding thus : Κυψέλῳ ᴋᴊ τοῖς προγόνοις EK ΓΟΝΟΥΣΣΗΣ ἦν γένος ἐξαρχῆς τῆς Σικυῶνος, ᴋᴊ πρό-γονος σφίσιν ἦν Μέλας ὁ Ἀντάσσε. " *The race of* Cypselus *and his* " *anceftors was originally from* " Gonuffa *in the country of* Si-" cyon, *and their progenitor was* " Melas,

" MELAS, *the fon of* ANTASSUS."
This agrees exactly with what
he fays in another place *, that
" MELAS, the fon of ANTASSUS,
came from GONUSSA beyond Si-
cyon, to ferve in the Dorian army
againſt Corinth, and that ALETES
with difficulty was prevailed upon
to receive him." The alteration
which I have propofed may feem
bold; but, I hope, it is not extravagantly fo. They who are
acquainted with MSS know, how
common it is for a word, or part
of a word, to be obliterated at
the beginning or end of a line.
Suppofing this to have happened
to the latter part of the word

* Corinthiac. p. 150.

Fo-

Γονϵσσης, the text would ſtand ϵκ Γον. Then comes a correc-tor, and puts the word Γονϵσσης in the margin, which the next tranſcriber inſerts in an impro-per place, changing EK ΓON at the ſame time into EKTON.

I cannot diſmiſs this argument without obſerving, that Sir Isaac, who would here make Aletes only *ſix* generations older than Cypselus, and has urged this as a fa&t that overturns the common Chronology; yet himſelf, in ano-ther place *, reckons up by name the ſucceſſors and lineal de-ſcendents of Aletes to the num-

* Chronol. p. 142.

ber

ber of *eight*, and adds to them 42 annual Archons, all intervening between ALETES and CYPSELUS. So little are the beſt and wiſeſt of men upon their guard, when they have a favourite opinion to ſupport.

I now proceed in the fifth place to conſider the difficulty ſuggeſted by the age of TERPANDER, the famous muſician. "*Athenæus,*" to uſe the words of Sir ISAAC [*], " tells us out of ancient " authors (*Hellanicus, Soſimus,* " and *Hieronymus)* that *Lycur-* " *gus* the Legiſlator was contem- " porary to *Terpander* the mu-

[*] Chronol. p. 58.

O 3 " ſician,

" fician, and that *Terpander* was
" the firſt man who got the vic-
" tory in the *Carnea*, in a ſo-
" lemnity of muſic inſtituted in
" thoſe feſtiyals in the 26th
" Olympiad." The inference is
plain, that if LYCURGUS lived till
the 26th Olympiad, the preceding Olympiads muſt be fictitious.
But here I muſt take the liberty
to ſay, that the words of Sir
ISAAC convey a falſe idea, much
too favourable to his ſyſtem. The
natural meaning of them is, that
there are three ancient hiſtorians,
who expreſsly give teſtimony
againſt the common notion of
LYCURGUS having lived a hundred years prior to the firſt Olympiad ;

piad ; and inftead of that bring
him a hundred years below it.
But the fact is, that not one of
thefe ancient authors fays either
that, or any thing like it. The
words of Athenæus literally tranf-
lated, run thus *. " That TER-
" PANDER was older than ANA-
" CREON is plain from the fol-
" lowing teftimonies. TERPAN-
" DER was the firft who obtained
" the prize in the *Carnea*, as
" HELLANICUS relates, both in
" his metrical and profe account
" of the *Carneonicæ.* Now the
" inftitution of the *Carnea* was
" in the 26th Olympiad, as *So-*
" *fimus* affirms in his Chronolo-

* L. XIV. p. 635.

O 4   " gy.

" gy. But HIERONYMUS, in his
" book upon *Citharoedi*, which
" is the fifth of his work upon
" *Poets*, fays, that TERPANDER
" lived in the time of LYCURGUS
" the Lawgiver, whom all wri-
" ters unanimoufly allow to have
" affifted IPHITUS the Elean in
" that inftitution of the Olympic
" games, which is reckoned the
" firft." We fee here three different writers, attefting three feparate unconnected facts, plainly
confidered by ATHENÆUS as
contradictory one to another, but
which muft be all brought together, and all fuppofed to be
true, before Sir ISAAC's conclufion can be made out from them.
                            This

This furely is very different from having each of the three witneffes fpeak to all the three facts. Even HIERONYMUS, the only one of them who brings LYCURGUS and TERPANDER together, agreed with all other writers, I fuppofe, (fince all writers, according to ATHENÆUS, were agreed) that LYCURGUS lived at the time of the firft inftitution of the Olympiads by IPHITUS, and therefore, had Hieronymus known the affertions of the other two, he would certainly have rejected one or other of them. So that here we have a point made out, it feems, not by three unanimous witneffes in the common way; but by three

who

who contradict and refute one
another.

What ground Hieronymus
might have for his fingular notion, that LYCURGUS and TER-
PANDER were contemporaries, it
is impoffible (ATHENÆUS having
faid nothing) to conjecture. But
fingular it certainly was ; for
PLUTARCH, in the Book *de Mufi-
ca* *, where he enquires very mi-
nutely into the age of TERPAN-
DER, does not give the leaft hint
of any fuch opinion having been
ftarted. He appears to have
placed him nearly where the
Arundel Marble places him, a

---

* Moral. p. 1132.

little

little before ARCHILOCHUS;
though the ſtory which he tells,
of his carrying away the prize
four times ſucceſſively at the Pythian Games is not conſiſtent
with that marble : unleſs we ſuppoſe him to mean Pythian Games,
celebrated at irregular intervals,
in ſome period prior to their final
eſtabliſhment *.

The ſame LYCURGUS furniſhes
Sir ISAAC with a ſixth objeƈtion
to the Olympic Chronology †,
which it is much eaſier to an-
ſwer, than to preſerve in anſwer-
ing the temper and decorum due

* Compare alſo Clem. Alex. Lib. I.
p. 333. Ed. Par.
† Chronol. p. 58.

to

to fo high a character. It feems there was a Difc at Olympia, which had the name of LYCUR-GUS infcribed upon it. Hence ARISTOTLE, according to PLU-TARCH *, inferred, that LYCUR-GUS was contemporary with Iphi-tus, the founder of the Olympic Games, and affifted in the eftab-lifhment of them. Sir ISAAC taking it for granted that this Difc was one of thofe ufed by the Athletes, finding that the Difc was a part of the *Pentathlos*, and having learnt from PAUSA-NIAS †, that from the time the Olympiads were celebrated in a regular feries, the *Pentathlos* was

* V. Lycurg, p. 85. Ed. Bryan.
† Eliac. I. p. 394.

never

never practifed till the 18th of thofe Olympiads, he concludes, that it was at this very time Ly-curgus was prefent, and confe-quently that his age has been fet near 140 years too high by the Chronologers. The conclufion is much too hafty, even though the premifes had been true. I need not inform the reader, that the invention and cuftom of cafting the Difc, as a trial of fkill, was older than Lycurgus, being mentioned by Homer. And though Homer himfelf, partak-ing of the common fate of his countrymen, lofes fomewhat of his antiquity in the hands of Sir Isaac, he is ftill admitted to have

I   been

been prior to LYCURGUS.. Now
the game of the Difc being con-
feffedly older than the *Pentathlos*,
what abfurdity is there in fup-
pofing that it might be feparate-
ly practifed at the Olympic Fefti-
val, as PINDAR exprefsly affirms
it was in fome places *, before it
was combined with other exer-
cifes to make the *Pentathlos?* It
is only in organized productions,
whether animal or vegetable, that
a part cannot exift before the
whole, but in civil inftitutions
nothing is more common. And
when it is fuppofed, that the
Game of the Difc could not exift
before the *Quinquertium*, it might

---

* Ifthm. I.

with

with equal juftice be faid, that the union of any two unconnect-ed offices in the fame perfon is a proof, that neither of them at any former period had been feparate-ly exercifed.

This would be a fufficient an-fwer. But the reader will be furprized to hear, that the fact is, not what Sir ISAAC has ftated it to be, but exactly the reverfe. PAUSANIAS does not fay, that the *Pentathlos*, or combination of the Difc, with four other Games, was firft practifed, or inftituted in the 18th Olympiad. His ac-count is to this effect *.   " After

* Eliac. I. p. 394.

" IPHITUS

" IPHITUS had revived the festi-
" val in the manner above re-
" lated, the memory of many
" antient customs was still lost,
" and it was by slow degrees that
" men came to the remembrance
" of them, and added to the
" Games whatever they happened
" to recollect. This is manifest.
" For reckoning from the time
" when the memorials of the
" Olympiads go on without in-
" terruption, the first prize given
" was for the foot-race, which
" was won by COROEBUS the Ele-
" an. Afterwards in the four-
" teenth Olympiad the *Diaulos*
" was added, and HYPENUS of
" Pisa carried away the olive-
" branch

" branch for the *Diaulos*, as
" ACANTHUS did in the next
" Olympiad. Then in the 18th
" Olympiad they recollected the
" *Pentathlos* and the wreftling."
The *Pentathlos* therefore was no
new invention of that time, but
the very words of PAUSANIAS
fhew, that it had been practifed
long before the revival of the
Olympic games by IPHITUS, fo
long indeed as to have gone into
difufe and oblivion. And hence
it follows, not only that the
Difc and *Pentathlos* might be as
old as the time at which LYCUR-
GUS is commonly placed; but,
if Sir ISAAC's inference be juft,
that the Difc was given by LY-

curgus, at the firſt inſtitution of the *Pentathlos*, it will lead alſo to another very unexpected concluſion, that Lycurgus himſelf muſt have lived a generation or two before the firſt Olympiad.

But what if the Diſc of Lycurgus, after all, ſhould have no relation to the *Pentathlos*, or to the Diſc thrown by the Athletes ? Pausanias informs us *, that there was preſerved at Elis a Diſc of Iphitus, on which was inſcribed the armiſtice proclaimed by the Eleans, the inſcription being not in a ſtrait line, but running circularly round the

* Eliac. I. p. 427.

Diſc.

Difc. Mr. JOHN JACKSON, in his Chronology, conjectures with great probability, that this was the Difc alluded to by PLUTARCH * ; and his opinion feems to be confirmed by the inference that ARISTOTLE drew from it, which was, not that LYCURGUS was the companion of IPHITUS in reftoring the Olympic Games, as Sir ISAAC furmifes, but that he affifted in fettling the armiftice.

The laft argument of Sir ISAAC againft the Olympic Chronology is taken from the lift of the Macedonian kings : and this indeed his manner of ftating it has ren-

* See before, p. 204.

P 2   dered

dered rather ſtrong. The interval of time which theſe kings are
to fill up terminates in two epochas, one undiſputed and certain,
the expedition of XERXES ; the
other much leſs determinate, the
reign of PHIDON *, the king of
Argos and invader of Elis. This
invaſion, according to PAUSA-
NIAS, who ſpeaks without any
marks of doubt or heſitation,
happened in the eighth Olympiad † ; but the Arundel marble,
if underſtood of the ſame PHI-
DON, carries him a full hundred
years higher. The latter computation was evidently the moſt
to Sir ISAAC NEWTON's purpoſe,

* See before, p. 179.
† Eliac. II. p. 509.

and

and therefore when he reasons fluently from this, without taking any notice of the other, he certainly does full justice to his argument. In a matter of such high antiquity, where most of the authors, who might have assisted us, are lost, I should think that I had as good a right to follow the authority of PAUSANIAS, even if it were single and unsupported, as Sir ISAAC has to follow the marble. But this is not the case. For STRABO, who was certainly well acquainted with the old Greek Historians, makes PHIDON the tenth from TEMENUS *, which exactly falls

* Lib. VIII. p. 549.

P 3   in

in with the reckoning of PAU-
SANIAS. PHIDON therefore being
fuppofed to reign in the eighth
Olympiad, the interval between
the end of that Olympiad and the
beginning of the 75th is exactly
264 years.

The number of Macedonian
princes who are to fill up that
interval, is a ftill more difputa-
ble point. HERODOTUS * makes
ALEXANDER, the contemporary of
XERXES, the feventh king from
the beginning of the monarchy;
and with this computation THU-
CYDIDES in effect agrees †. In
the lift given by HERODOTUS,

* Lib. VIII. c. 139.
† Lib. II. c. 100.

PER-

PERDICCAS ſtands firſt : which, I apprehend, is no further true than that PERDICCAS was the firſt who reigned under the title of King : which is preciſely what SOLINUS aſſerts *. But if we may believe other ancient authors, PERDICCAS was by no means the perſon, or contemporary with the perſon, who under the reign of PHIDON quitted Argos, and removed into Macedonia. This perſon, by every author but HERODOTUS, is called CARANUS, whom we learn from SYNCELLUS to have been the brother of PHIDON †. He is mentioned alſo by PLUTARCH ‡,

* Cap. IX.
† Syncell. Chron. p. 158. Ed. Venet.
‡ Vit. Alexand. p. 6.

P 4   PAU-

PAUSANIAS *, and DIODORUS SI-
CULUS †; by SATYRUS, an ancient
author quoted by THEOPHILUS
bifhop of Antioch ‡; and among
the Latins by LIVY §, PATERCU-
LUS ‖, JUSTIN **, and SOLI-
NUS ††. Then follows another
queftion, whether any generations intervened between CARA-
NUS and PERDICCAS. The lift
of the Macedonian Kings in SYN-
CELLUS ‡‡ inferts two, by the
names of COENUS and TYRIM-
MAS ; and he alfo informs us,

* Boeot. p. 794.
† Ap. Syncell, p. 209.
‡ Ad Autolyc. II. p. 96. Ed. Wolf.
§ Dec. V. Lib. V.
‖ Lib. I.
** Lib. VII.
†† Cap IX.
‡‡ Chron. p. 209. Ed. Venet.

that

that Coenus was the fon of Caranus, and Tyrimmas of Coenus. In other refpects his lift
agrees exactly with that of Herodotus, which I hope will be
no diminution of its authority.
The above quoted Satyrus alfo
inferts the fame two names between Caranus and Perdiccas;
though he differs from Herodotus and Syncellus, by leaving
out Argæus, the fon of Perdiccas; whofe exiftence however
is eftablifhed both by the authority of Justin *, and by medals †.
Thefe different omiffions are eafily accounted for from the negligence of tranfcribers in writing

* l. c.　　　† Not. ad Herodot. l. c.

out

out a tedious genealogy. It would not be ſo eaſy to account for their inſerting a name without authority; becauſe this would not be negligence but invention. In the generations ſubſequent to PERDICCAS, we find ALCETAS, the father of AMYNTAS, omitted by JUSTIN. Yet there can be no doubt, I think, of his having been really the ſon of AEROPUS, and father of AMYNTAS, becauſe, as he is placed only three generations before HERODOTUS, it is hardly poſſible that he could have been inſerted by miſtake, and the other two liſts concur in retaining him.

There

There is upon the whole then indisputable evidence, that CA-RANUS was the person who removed from Argos, and laid the first foundation of the Macedonian kingdom. There is also good authority for supposing that PERDICCAS, who completed the work of CARANUS, and first assumed the title of King, was not the *brother*, as Sir ISAAC from the ambiguous authority of HE-RODOTUS presumes, but the *great grandson* of CARANUS. The list and order of generations will therefore stand thus, precisely as in SYNCELLUS: 1. CARANUS. 2. COENUS. 3. TYRIMMAS. 4. PER-

4. PERDICCAS.   5. ARGÆUS.

6. PHILIPPUS.   7. AEROPUS.

8. ALCETAS.   9. AMYNTAS.

10. ALEXANDER.   Ten kings make nine generations, as it is always neceſſary to ſtrike off either the firſt or the laſt of the liſt. Divide then 264, the number of years between Phidon and the expedition of Xerxes, by 9, the number of generations, and the quotient will be 29, with a remainder of 3 : that is, the portion of time for each generation will be exactly 29 years and 4 months; which is conſiderably leſs than Sir ISAAC himſelf allows. And I have already ſhewn at large, that if we calculate at all,

it

it muſt be by generations, the number of kings furniſhing no ground whatever for rational argument.

And now having, I think, evinced the inſufficiency of Sir IsAAC's arguments to overthrow the Chronology of the Olympiads, I will, to ſhew my own fairneſs, produce two which he has overlooked. The Scholiaſt of PINDAR, in his Commentary on the ſecond and third Olympic Odes, gives us, as it ſhould ſeem, the genealogy of THERON, the contemporary of PINDAR. In the former place he begins it thus * : 1. LAIUS.   2. OEDIPUS,

* Schol. in Pindar, O. II. v. 82.

3. Po-

3. POLYNICES. 4. THERSANDER. 5. TISAMENUS. 6. ANTESION. 7. THERAS. 8. SAMUS. 9. TELEMACHUS, who removed from the ifland of Thera, and fettled in Sicily. 10. CHALCIOPEUS. 11. ÆNESIDAMUS. 12. THERON. In the latter he gives only the immediate progenitors of Theron * : 1. TELEMACHUS, who depofed the tyrant PHALARIS. 2. EMMENIDES. 3. ÆNESIDAMUS. 4. THERON. THERAS, the feventh in the firft lift, was contemporary with TEMENUS, the conqueror of Argos; from whom, according to this lift, THERON would only be the fifth

* O. III. v. 68.

in

in defcent. Now PINDAR, having been born in Olympiad
LXV. we cannot fuppofe THE-
RON, whom he celebrates, to
have been born earlier than
Olympiad LV. But from TE-
MENUS to the beginning only of
the Olympiads were eight generations. So that if the Scholiaft
has given us the compleat genealogy, it will follow that there
were no lefs than 55 fictitious
Olympiads; which if any one
is difpofed to believe, I will not
be his hindrance.

A fecond argument might be
brought from PAUSANIAS, who
tells us *, that PYTHAGORAS the

* Corinthiac. p. 140.

phi-

philofopher was the great-grand-
fon of HIPPASUS, and that HIP-
PASUS was contemporary with
REGNIDAS, the grandfon of TE-
MENUS. This makes a ftill greater
defalcation of time ; and I fhall
therefore leave it in full force,
that the advocates for this part of
Sir ISAAC NEWTON's Syftem of
Chronology may difpofe of it as
they pleafe.

Before I conclude, I have one
general remark to make upon Sir
ISAAC's book ; that he finds fault
with the earlier part of the Græcian Hiftory for having no Chronology ; and yet fuppofes, that
when Chronology, that is, technical

nical Chronology, was introduced by TIMÆUS and others,
the only ufe made of it was to
falfify their hiftory. This makes
it neceffary to explain, in a few
words, my notion what Chronology is, and what it is not. I
fay then, that the Genealogy of
a particular family, a feries of
kings or prieftesses, a lift of archons, or the records of a public folemnity like the Olympic
Games ; none of thefe are Chronology. But Chronology is that
fcience, which compares thofe
lifts, genealogies, and records,
together, and adjufts them one
to another ; making, if poffible,
one confiftent whole. This is a

Q   work

work that requires, no doubt, the hand of a mafter, and it requires alfo an unprejudiced mind. For if the chronologer has any favourite point to eftablifh, if, for inftance, he is defirous of extending or contracting any particular period, he will be tempted in his account of public tranfactions to imitate PROCRUSTES ; to mutilate or ftretch them out, as may beft ferve his purpofe. I do not know that the antient chronologers were under any fuch temptation ; that there was any particular fyftem of hiftory, which they were obliged at all events to make good ; and therefore whatever

errors

errors they may have committed, I prefume they were only errors of judgement. With refpect to the genealogies and records, which preceded this technical chronology, they are ftill further removed from any fufpicion of infidelity. The compilers of them purfued no fyftem, and therefore could be mifled by none. The want of Chronology therefore, with which Sir Isaac reproaches the older Greeks, is a circumftance which, in another point of view, may be confidered as ftrongly fupporting their credit.

POST-

POSTSCRIPT.

IN writing p. 179 I overlooked a very material circumstance, mentioned by PAUSANIAS *, and confirmed by STRABO †; which is, that the Eleans made no entry of the Olympiad at which PHIDON the Argive presided. Now PAUSANIAS, having inspected the record at Olympia, could not be misinformed with respect to the particular Olympiad. And hence the age of PHIDON is fixed to a certainty. He could neither be so ancient as the marble makes him, nor so modern as he is supposed by Sir ISAAC NEWTON.

* Eliac. II. p. 509.
† L. VIII. p. 549.

The

The paſſage of STRABO proves alſo the truth of what I have aſ-ſerted from PAUSANIAS *, that the account kept by the Eleans of the Olympic Games was in the nature of a record, and that it was kept from the time that the Olympiads are referred to numerically.

How perfect this record was, and how carefully PAUSANIAS had examined it, might be further proved by what he ſays of XENODAMUS of Anticyra, upon whoſe ſtatue there was an in-ſcription, importing, that he had

* See p. 124—128.

ob-

obtained a victory in the *Pancra-tium* at the Olympic Games. " If this inscription be true," says he *, " it should seem, that XENODAMUS obtained this prize in the 211th Olympiad. For this is the only Olympiad omitted in the register of the Eleans." It is plain therefore, that PAUSANIAS had inspected the register from beginning to end, and that he had found it perfect in every instance but this. The reason of this particular Olympiad being omitted was probably on account of the irregular interference of NERO, who was present at it.

* Phocic. p. 892.

The

The correction of HERODOTUS proposed in p. 184. is countenanced by a similar passage of PAUSANIAS, Arcad. p. 631. Δῆλα ἓν ἐςι Χαλκώδονἶα, ὃ τὸν ἐξ Εὐϐοίας, ἢ Τελαμῶνα, ὃ τὸν Αἰγινήτην, ἐπὶ Ἠλέὃς Ἡρακλἕ μετεσχηκέναι τῆς ςρατιᾶς.

T H E   E N D.

www.ingramcontent.com/pod-product-compliance
Lightning Source LLC
Chambersburg PA
CBHW021521210326
41599CB00012B/1333